Mastering iOS 18 De

Take your iOS development experience to the next level with iOS, Xcode, Swift, and SwiftUI

Avi Tsadok

Mastering iOS 18 Development

Copyright © 2024 Packt Publishing

All rights reserved. No part of this book may be reproduced, stored in a retrieval system, or transmitted in any form or by any means, without the prior written permission of the publisher, except in the case of brief quotations embedded in critical articles or reviews.

The author acknowledges the use of cutting-edge AI, in this case Grammarly, with the sole aim of enhancing the language and clarity within the book, thereby ensuring a smooth reading experience for readers. It's important to note that the content itself has been crafted by the author and edited by a professional publishing team.

Every effort has been made in the preparation of this book to ensure the accuracy of the information presented. However, the information contained in this book is sold without warranty, either express or implied. Neither the author, nor Packt Publishing or its dealers and distributors, will be held liable for any damages caused or alleged to have been caused directly or indirectly by this book.

Packt Publishing has endeavored to provide trademark information about all of the companies and products mentioned in this book by the appropriate use of capitals. However, Packt Publishing cannot guarantee the accuracy of this information.

Group Product Manager: Rohit Rajkumar
Publishing Product Manager: Chayan Majumdar
Book Project Manager: Sonam Pandey
Senior Editor: Rashi Dubey
Technical Editor: K Bimala Singha
Copy Editor: Safis Editing
Indexer: Subalakshmi Govindhan
Production Designer: Ponraj Dhandapani
DevRel Marketing Coordinator: Nivedita Pandey

First published: November 2024

Production reference: 1091024

Published by Packt Publishing Ltd.
Grosvenor House
11 St Paul's Square
Birmingham
B3 1RB, UK

ISBN 978-1-83546-810-4

www.packtpub.com

This book was a year in the making, filled with countless hours of research, investigation, and a few too many late-night coffee runs. While I loved every minute of it, I have to give a big shoutout to my amazing wife, Tammy, for her endless support and understanding, and to my kids, Harel and Maya, who heroically endured my "just five more minutes" promises. And last but not least, a huge thank you to my trusty alarm clock, which somehow managed to drag me out of bed at 5 A.M. every day to pursue my passion for writing.

– Avi Tsadok

Contributors

About the author

Avi Tsadok, a seasoned iOS developer with a 14-year career, has proven his expertise by leading projects for notable companies, such as Any.do, a top productivity app. He is currently at Melio Payments, where he steers the mobile team. Known for his ability to simplify complex tech concepts, Avi has written 4 books and published 40+ tutorials and articles that enlighten and empower aspiring iOS developers. His voice resonates beyond the page, as he's a recognized public speaker and has conducted numerous interviews with fellow iOS professionals, furthering the field's discourse and development.

It was a challenging year for me and my family, and writing this book became a source of strength during those difficult times. I want to express my deepest gratitude to Packt for their unwavering support and professionalism, with special thanks to Sonam, Chayan, and Rashi – your efforts did not go unnoticed. Finally, a heartfelt thank you to my family – Tammy, Harel, and Maya – who gave me the strength to complete this incredible project.

About the reviewers

Hritik Raj is an iOS developer-turned-product manager at meShare, with extensive experience working on IoT apps used by millions of users. Passionate about learning, Hritik taught himself SwiftUI through books and tutorials during his time at the University of Illinois at Urbana-Champaign, where he graduated with a degree in computer engineering. His expertise spans both development and product management, giving him a unique perspective on building user-centric apps.

Ruy de Ascenção Neto is an alumnus of the Apple Developer Academy in Brazil with six years of experience in Objective-C, Swift, and SwiftUI. He has experience in developing applications for iPhone, Apple TV, and Apple Watch. He has worked at both national and international banks, as well as streaming companies.

Table of Contents

Preface

Part 1: Getting Started with iOS 18 Development

1

What's New in iOS 18   3

Technical requirements	4	Having more control over scroll views	13
Understanding iOS 18 background	4	Observing the scroll view position	13
Introducing Swift Testing	5	Observing items' visibility	15
Introducing Swift Data Improvements	5	Changing the text rendering behavior	16
Unique value	6	Positioning sub-views from another view	17
History API	6		
Custom data stores in Swift Data	6	Entering the AI revolution	19
Introducing zoom transition	9	Summary	20
Adding a floating tab bar	11		

2

Simplifying Our Entities with SwiftData   21

Technical requirements	21	Adding the @Attribute macro	30
Understanding SwiftData's background	22	Going non-persistent with transient	32
Defining a SwiftData model	23	Exploring the container	33
Expanding the @Model macro	23	Setting up ModelContainer	33
Adding relationships	26	Connecting the container using the modelContainer modifier	34
SwiftData relationship deletion rules	27		
Defining the inverse relationship	28	Working with ModelConfiguration	35

Fetching and manipulating our data using model context	36	Learning the basic migration process	42
		Creating a version schema	43
Saving new objects	37	Creating the migration stages and plan	44
Fetching objects	38	Connecting the migration plan to our container	46
Migrating our data to a new schema	42	Summary	47

3

Understanding SwiftUI Observation 49

Technical requirements	49	Excluding properties from observation using @ObservationIgnored	55
Going over the SwiftUI observation system	50	Observing computed variables	56
Conforming to the ObservableObject protocol	50	Working with environment variables	59
Explaining the problem with the current observation situation	51	Adding an environment variable by type	59
		Adding environment variable by key	61
Adding the @Observable macro	52	Binding objects using @Bindable	63
Learning how the @Observable macro works	53	Migrating to Observable	65
		Summary	66

4

Advanced Navigation with SwiftUI 67

Technical requirements	67	Understanding the Coordinator's principles	77
Understating why SwiftUI navigation is a challenge	68	Building the Coordinator object	79
		Adding CoordinatorView	80
Exploring NavigationStack	68	Calling the coordinator straight from the view	81
Separating the navigation destination using the navigationDestination view modifier	69	Navigating with columns with NavigationSplitView	82
Using data models to trigger navigation	71	Creating NavigationSplitView	83
Responding to the path variable	73	Moving to three columns	85
Working with different types of data using NavigationPath	75	Summary	87
Working with the Coordinator pattern	77		

5

Enhancing iOS Applications with WidgetKit — 89

Technical requirements	89	Customize our widget	104
The idea of widgets	90	Using the AppEntity in our Widget	107
Understanding how widgets work	91	**Keeping our widgets up to date**	**108**
Adding a widget	91	Reload widgets using the WidgetCenter	108
Configuring our widget	93	Go to the network for updates	110
Working with static configuration	95	**Interacting with our widget**	**111**
Understanding the Timeline Provider for Widgets	95	Opening a specific screen using links	111
		Adding interactive capabilities	112
Building our widget UI	100	**Adding a control widget**	**114**
Working with timeline entries	101	**Summary**	**117**
Adding animations	102		

6

SwiftUI Animations and SF Symbols — 119

Technical requirements	119	**Performing advanced animations**	**125**
The importance of animations	120	Performing transitions	125
Understanding the concept of SwiftUI animations	120	Executing keyframe animations	129
		Animating SF Symbols	**133**
Performing basic animations	121	Modifying symbol colors	135
Using the animation view modifier	121	Localizing our symbols	137
Using the withAnimation function	122	**Summary**	**138**
Bringing some life to our animations with spring animations	124		

7

Improving Feature Exploration with TipKit — 139

Technical requirements	140	What do tips look like?	142
Learning the importance of tips	140	Adding our first tip	143
Understanding the basics of TipKit	140	Dismissing tips	147
		Defining the tip ID	148

Customizing our tips	149	Grouping tips with TipGroup	159
Customizing our tips' appearance	149	**Customizing display frequency**	**161**
Adding actions	152	Setting the max display count for a specific tip	161
Adding tips rules	**154**	Setting our tips' display frequency	161
Adding a rule based on a state	154	**Summary**	**162**
Adding a rule based on events	157		

8

Connecting and Fetching Data from the Network 163

Technical requirements	163	Integrating network calls within app flows	175
Understanding mobile networking	164	Just-in-time fetching	175
Handling an HTTP request	165	Read-through cache	176
Basic HTTP request methods	165	Incremental loading	178
Working with URLSession	166	Full data sync with delta updates	179
Handling the response	167	**Exploring Networking and Combine**	**182**
		Summary	**185**

9

Creating Dynamic Graphs with Swift Charts 187

Technical requirements	187	Visualizing functions with Charts	206
Why charts?	188	Allowing interaction using ChartProxy	208
Introducing the Swift Charts framework	188	Adding an overlay to our chart	209
Creating charts	190	Responding to the user's gesture	211
Creating BarMark chart	190	Finding the closest data point to the user's touch	211
Creating LineMark charts	196	**Conforming to the Plottable protocol**	**213**
Creating a SectorMark chart	200	**Summary**	**214**
Creating an AreaMark chart	202		
Creating a PointMark chart	204		

Part 2: Refine your iOS Development with Advanced Techniques

10

Swift Macros 217

Technical requirements	217	Examining our Swift Macros package structure	229
What is a Swift macro?	218	Declaring our macro	230
Exploring SwiftSyntax	219	Implementing the macro	232
Parsing and AST	220	**Handling macros errors**	**238**
Setting up SwiftSyntax	220	**Adding tests**	**240**
Building our Abstract Syntax Tree	222	**Practice exercises**	**243**
Creating our first Swift macro	**227**	**Summary**	**243**
Adding a new Swift macro	227		

11

Creating Pipelines with Combine 245

Technical requirements	245	Connecting the custom publisher and subscriber	258
Why use Combine?	246	Working with operators	258
Going over the basics	247		
Starting with the publisher	247	**Learning about Combine using examples**	**264**
Setting up the subscriber	248	Managing UIKit-based view state in a view model	264
Connecting operators	249	Performing searches from multiple sources	265
Delving into Combine components	**250**	Validating forms	267
Creating a custom publisher	251	**Summary**	**269**
Working with Subjects	252		
Creating a custom subscriber	255		

12

Being Smart with Apple Intelligence and ML — 271

Technical requirements	272	Classifying audio using the Sound Analysis framework	282
Going over the basics of AI and machine learning	272	Performing a semantic search with Core Spotlight	284
Learning the differences between AI and machine learning	272	**Integrating custom models using CoreML**	**288**
Delving into the ML model	273	Getting to know the Create ML application	289
Training the model	273	Building our Spam Classifier model	291
Apple intelligence and ML	**274**	Using our model with Core ML	298
Exploring built-in ML frameworks	**274**	Where to go from here	300
Interpreting text using NLP	275	**Summary**	**300**
Analyzing images using the Vision framework	279		

13

Exposing Your App to Siri with App Intents — 301

Technical requirements	301	Conforming to AppEntity	312
Understanding the App Intents concept	302	Creating an Open a task intent	314
		Chaining app intents	315
Creating a simple app intent	302	Integrating our intent to other intents	316
Running the intent with the Shortcuts app	303	**Adjusting our app intents to work with Apple Intelligence**	**320**
Creating an app shortcut	304		
Adding a parameter to our app intent	305	Exploring the Assistant Schema	320
Returning a custom view	307	Creating AssistantEntity	323
Having multiple result types	309	**Summary**	**326**
Adding confirmation and conditions	310		
Formalizing our content using app entities	**312**		

14

Improving the App Quality with Swift Testing — 327

Technical requirements	327	Managing our tests	340
Understanding the importance of testing	328	Going over the testing structure	341
		Grouping our test functions into test suites	342
Learning the testing history in Apple platforms	329	Building test plans	344
		Setting up a Scheme	349
Exploring the Swift Testing basics	330	Tips to write testable code	350
Adding a basic test	331	Writing pure functions	351
Providing names to our test functions	334	Separating your code based on concerns	351
Enabling and disabling tests	334	Performing mocking using protocols	352
Tagging our test functions	336	Summary	354
Working with arguments	338		

15

Exploring Architectures for iOS — 355

Technical requirements	355	Combining the multi-layer architecture with modules	365
Understanding the importance of architecture	355	Building hexagonal architecture	367
Learning what exactly architecture is	357	Comparing the different architectures	374
Going over the different architectures	358	By separation of concerns	374
Separating our project into layers	358	By testing	375
Separating our project into modules	364	By maintenance and scalability	375
		Summary	377

Index — 379

Other Books You May Enjoy — 392

Preface

Before we begin our journey, let me welcome you to iOS 18 development!

Looking back at 2008 and trying to recall how the iOS SDK was then, I'm amazed at how it has developed and evolved over the years. Back then, all we had to know as iOS developers was how to create a `UITableView`, add some buttons, and be an expert in a fantastic design pattern called MVC. That was enough to make a standard app and even get hired as an iOS developer.

But we are not in 2008, and things have changed a bit—well, maybe more than just a bit. What has changed? Everything! The programming language, the UI framework, the design patterns, and even the IDE. But it's not only what has changed but also what was added.

In 2008, the iOS SDK (previously known as the iPhone SDK) contained less than 25 frameworks. Now, we have over 200 frameworks—that's nearly ten times more!

We have frameworks for animation, gaming, testing, machine learning and AI, security, data, and many more. These days, being an iOS developer is much more than adding a list and a button—it is understanding the capabilities of the iOS SDK and choosing the right approach and technology.

And that's precisely the book's goal. It is not a reference or documentation of Apple's technology – you can get that online, and it's probably more up to date. This book is a window into what the iOS SDK is capable of so you can improve your development skills even further.

This book's information was carefully selected to cover the most exciting and valuable parts of modern iOS development, including a persistent store, testing, advanced SwiftUI concepts, networking, macros, architectures, and even machine learning and AI. It is impossible to cover everything, and that's not the intention.

However, understanding the topics in this book will help you get the most out of the iOS SDK.

Who this book is for

The book is not for starters! Developers who read it must have a basic knowledge of Swift, SwiftUI, Xcode, and basic concepts in iOS development, such as animation, networking, and persistent data. So, this is not a "get started with iOS development" book – I assume you have written a few lines in Swift and created some great UI in SwiftUI.

Three primary personas are the target audience of this book:

- iOS senior developers who want to stay up to date with the latest Apple technologies
- iOS team tech leads who want to leverage their team skills
- Mid-level developers who want to step up to the senior's area

What this book covers

Chapter 1, What's New in iOS 18, provides an overview of iOS 18 and covers the SDK's changes and new capabilities. It also discusses the iOS 18 approach and the different trends.

Chapter 2, Simplifying Our Entities with SwiftData, covers a new framework from Apple that replaces Core Data for having persistent storage. This chapter covers everything from setup, performing operations, queries, and migration.

Chapter 3, Understanding SwiftUI Observation, provides an overview of a critical aspect of SwiftUI. This chapter discusses Apple's new observation framework, restructures our understanding of the different property wrappers' roles, and dives deep into how they work underneath.

Chapter 4, Advanced Navigation with SwiftUI, covers another massive topic in iOS. It discusses the complexity of SwiftUI navigation and provides real-world examples and patterns for handling it.

Chapter 5, Enhancing iOS Applications with WidgetKit, explains the idea of widgets; covers how to add, maintain, and design widgets; and covers the new capabilities of widgets, such as interactions and control widgets in iOS 18.

Chapter 6, SwiftUI Animations and SF Symbols, will help our app be more delightful and engaging. We will understand the importance of animations and their concept in SwiftUI, perform basic animations, and animate SF Symbols.

Chapter 7, Improving Feature Exploration with TipKit, discusses an interesting SDK that bridges the gap between the developer and product perspectives. We will learn how to add tips to our app, design them, and control their appearance rules.

Chapter 8, Connecting and Fetching Data from the Network, touches on one of the most essential topics in iOS: retrieving data from the network. We will understand how to handle HTTP requests and connect the Combine framework to our flows.

Chapter 9, Creating Dynamic Graphs with Swift Charts, is the most colorful chapter in this book. We will learn about the different types of charts available, create different charts, and even implement user interactions so our users can gain even more value.

Chapter 10, Swift Macros, is an advanced chapter that covers a complex yet powerful topic. This chapter dives into the SwiftSyntax framework, which stands behind Swift Macros and helps us add and test our first Swift Macro. This topic becomes crucial as many frameworks' APIs are based on Swift Macros.

Chapter 11, Creating Pipelines with Combine, covers the fundamental concepts of declarative programming. The chapter discusses the Combine framework basics, delving into the different Combine components such as publishers, subscribers, and operators, and also provides tools to integrate Combine into real-life flows.

Chapter 12, Being Smart with Apple Intelligence and ML, explores the fascinating world of machine learning. We will review the basics of machine learning and AI and try the built-in machine learning frameworks in iOS, such as NLP, vision, and sound analysis. Not only that, the chapter also explains how to train our own model and use it in our apps.

Chapter 13, Exposing Your App to Siri with App Intents, takes our existing apps and exposes their capabilities, such as actions and contents, to Siri. This chapter provides a great way to prepare our apps for the AI era.

Chapter 14, Improving the App Quality with Swift Testing, touches on a critical but unpopular topic in iOS development. The new Swift Testing frameworks make testing more straightforward and more natural. We will set up a testing target, write our first test function, and understand how to manage the different test plans, suites, and configurations.

Chapter 15, Exploring Architectures for iOS, aims to explain the different architectural concepts and help you choose an exemplary architecture that can balance simplicity, scale, and maintainability over time. There's no point in having an excellent idea for a room if you don't know how to build your house, right?

To get the most out of this book

You will need to understand Apple's platform engineering—Xcode, Swift, SwiftUI—and have some experience writing an iOS app—having experience writing a simple screen is not enough for the book's content to be valuable.

Software/hardware covered in the book	Operating system requirements
Xcode	macOS
iOS SDK	
Create ML	

If you are using the digital version of this book, we advise you to type the code yourself or access the code from the book's GitHub repository (a link is available in the next section). Doing so will help you avoid any potential errors related to the copying and pasting of code.

Download the example code files

You can download the example code files for this book from GitHub at https://github.com/PacktPublishing/Mastering-iOS-18-Development.

If there's an update to the code, it will be updated in the GitHub repository.

We also have other code bundles from our rich catalog of books and videos available at https://github.com/PacktPublishing/. Check them out!

Conventions used

There are a number of text conventions used throughout this book.

`Code in text`: Indicates code words in text, database table names, folder names, filenames, file extensions, pathnames, dummy URLs, user input, and Twitter handles. Here is an example: "In `NavigationStack`, there's a new view modifier called `navigationDestination`, which allows us to define a destination separately according to a state change."

A block of code is set as follows:

```
enum Screen: Hashable {
    case signin
    case onboarding
    case mainScreen
    case settings
}

@State var path: [Screen] = []
```

When we wish to draw your attention to a particular part of a code block, the relevant lines or items are set in bold:

```
struct CoordinatorView: View {

    @ObservedObject private var coordinator = Coordinator()

    var body: some View {
        NavigationStack(path: $coordinator.path) {
            AlbumListView()
                .navigationDestination(for: PageAction.self, destination: { pageAction in
                    coordinator.buildView(forPageAction: pageAction)
                })
        }
    }
```

```
            .environmentObject(coordinator)
    }
}
```

Any command-line input or output is written as follows:

```
@testable import Chapter14
```

Bold: Indicates a new term, an important word, or words that you see onscreen. For instance, words in menus or dialog boxes appear in **bold**. Here is an example: "The code creates a blue circle and a button saying **Start**."

> **Tips or important notes**
> Appear like this.

Get in touch

Feedback from our readers is always welcome.

General feedback: If you have questions about any aspect of this book, email us at `customercare@packtpub.com` and mention the book title in the subject of your message.

Errata: Although we have taken every care to ensure the accuracy of our content, mistakes do happen. If you have found a mistake in this book, we would be grateful if you would report this to us. Please visit `www.packtpub.com/support/errata` and fill in the form.

Piracy: If you come across any illegal copies of our works in any form on the internet, we would be grateful if you would provide us with the location address or website name. Please contact us at `copyright@packt.com` with a link to the material.

If you are interested in becoming an author: If there is a topic that you have expertise in and you are interested in either writing or contributing to a book, please visit `authors.packtpub.com`.

Share Your Thoughts

Once you've read *Mastering iOS 18 Development*, we'd love to hear your thoughts! Scan the QR code below to go straight to the Amazon review page for this book and share your feedback.

`https://packt.link/r/1835468101`

Your review is important to us and the tech community and will help us make sure we're delivering excellent quality content.

Download a free PDF copy of this book

Thanks for purchasing this book!

Do you like to read on the go but are unable to carry your print books everywhere?

Is your eBook purchase not compatible with the device of your choice?

Don't worry, now with every Packt book you get a DRM-free PDF version of that book at no cost.

Read anywhere, any place, on any device. Search, copy, and paste code from your favorite technical books directly into your application.

The perks don't stop there, you can get exclusive access to discounts, newsletters, and great free content in your inbox daily

Follow these simple steps to get the benefits:

1. Scan the QR code or visit the link below

```
https://packt.link/free-ebook/9781835468104
```

2. Submit your proof of purchase
3. That's it! We'll send your free PDF and other benefits to your email directly

Part 1: Getting Started with iOS 18 Development

In this part, you will review all the new capabilities of iOS 18. We will explore exciting topics, such as SwiftData, Observation, and SwiftUI navigation. In addition, we will build widgets with WidgetKit, animate our views, add tips and graphs to our apps, and learn how to build a great network layer.

This part contains the following chapters:

- *Chapter 1, What's New in iOS 18*
- *Chapter 2, Simplifying Our Entities with SwiftData*
- *Chapter 3, Understanding SwiftUI Observation*
- *Chapter 4, Advanced Navigation with SwiftUI*
- *Chapter 5, Enhancing iOS Applications with WidgetKit*
- *Chapter 6, SwiftUI Animations and SF Symbols*
- *Chapter 7, Improving Feature Exploration with TipKit*
- *Chapter 8, Connecting and Fetching Data from the Network*
- *Chapter 9, Creating Dynamic Graphs with Swift Charts*

1
What's New in iOS 18

Apple introduced iOS 18 in WWDC 2024 as part of its annual developer's conference, alongside macOS, tvOS, iPadOS, watchOS, and visionOS.

Utilizing our app's latest features and capabilities in each major OS release gives us a competitive advantage. Here are the reasons why Apple chose to improve particular domains in the SDK – market research or technology trends are good enough reasons to adopt new technologies.

However, to understand iOS 18 improvements, we first must understand the background for this version – that's one of this chapter's goals.

In this chapter, we will cover the following topics:

- Understanding iOS 18 background
- Exploring Swift Testing
- Learning about the new Swift Data improvements
- Trying the new zoom transition
- Adding a floating tab bar to our iPad apps
- Having more control over scroll views in SwiftUI
- Changing the text rendering behavior
- Positioning sub-views from another view
- Entering the AI revolution

If that sounds like an exciting chapter, you are not wrong. Let's start by understanding the background of iOS 18.

Technical requirements

For this chapter, it's essential to download Xcode version 16.0 or higher from the App Store.

Ensure that you're operating on the most recent version of macOS (Ventura or newer). Just search for Xcode in the App Store, choose the latest version, and proceed with the download. Open Xcode and complete any further setup instructions that appear. After Xcode is completely up and running, you can begin.

This chapter includes many code examples, and can be found in the following GitHub repository:

```
https://github.com/PacktPublishing/Mastering-iOS-18-Development/tree/main/Chapter%201
```

Understanding iOS 18 background

Releasing a major iOS version is always a big deal, even if it's the 18th already. Let's try to analyze the iOS SDK before iOS 18:

- **SwiftUI** is becoming more mature and capable. However, some features, such as complex animations or transitions, gesture handling, navigation, and drawing, remain challenging to implement using SwiftUI.
- **Core Data** is the go-to framework for most iOS developers as a solution for storing data persistently.
- While **XCTest** is considered a robust and convenient testing framework, it lacks features that are commonly available on other platforms, such as parameterized testing and better testing organization.
- **WidgetKit**'s popularity proves that the ability to show information at a glance is crucial in today's world.

No one can argue that this list is important. However, one critical topic that Apple didn't focus on until WWDC 2024 is artificial intelligence.

The rise of OpenAI's ChatGPT, followed by thousands of machine learning and AI tools, put Apple in a weird situation. This is not the first time Apple has left behind some temporary trend, but this time it was different. AI's potential influence on humanity indicates that this is not a regular trend or technology evolution; it is practically a revolution that will change the world.

The question is, where is Apple with its set of platforms and technologies? Does it have an answer to the AI revolution?

Before diving into that question, let's first review the new features and frameworks introduced in iOS 18 and explore how the latest version tackles some of the key challenges we face in iOS development. Don't worry, though – we'll cover the AI revolution in the final section and throughout the book. Now, let's discuss a new framework – Swift Testing.

Introducing Swift Testing

Swift Testing is a new framework with a new and refreshing approach to testing. Swift Testing contains modern features such as macros, which work with structs instead of classes and can tag tests and test suites.

Swift Testing is supposed to replace XCTest, which was introduced in 2013 as part of Xcode 5. XCTest belongs to older times when Objective-C was the dominant language. However, Swift took over, and Apple understood that iOS developers needed a modern testing framework.

Here's a simple test function:

```
@Test("Test view model increment function", .enabled(if: AppSettings.
CanDecrement), .tags(.critical))
func testViewModelIncrement() async throws {
//      preparation
        let viewmodel = CounterViewModel()
        viewmodel.count = 5

//      execution
        viewmodel.increment(by: 1)

//      verification
        #expect(viewmodel.count == 6)
    }.
```

We can see how simple it is to write a test function in Swift Testing. Notice the preceding Swift macro, which configures and tags the function as critical in addition to providing the test description.

If your app doesn't have a test function, Swift Testing is a great way to start (to read more about Swift Testing, go to *Chapter 14*).

Now, let's discuss another new framework that handles our persistent store.

Introducing Swift Data Improvements

Swift Data was introduced in WWDC 2023 as part of iOS 17, and its goal was to replace the old but popular Core Data framework.

Swift Data provides a modern API based on Swift, which can help reduce friction when working with persistent stores. One of the trends we see in Apple development tools is moving away from GUI to code-based tools. A good example is SwiftUI – even though it is possible to drag and drop components to build a user interface, the primary way to do this is in code. The same goes for App Intents and Swift Package Managers. The data layer goes through the same concept – in Swift Data, we don't have any data model editor, so we build our data model using only code.

For example, here's how to create a data model for a `Book` entity:

```
@Model
class Book {
    var author: String
    var title: String
    var publishedDate: Date
}
```

At first glance, it seems like a regular `Book` class – and it is! This time, we added the `@Model` macro, which does all that magic.

When Swift Data was introduced, it already had many features, such as relationships and deletion rules. Despite that, many developers felt that the framework wasn't mature enough to replace Core Data.

In iOS 18, Apple added some features to Swift Data that, if it is not already there, will bring it closer to where it should be.

Unique value

The first and maybe most important new feature in iOS 18 is the ability to construct a **unique value** for the model based on its attributes:

```
#Unique<Book>([\.name, \.publicationName])
```

In this case, the `Book` class's unique identifier is based on combining the `name` and `publicationName` attributes.

History API

Another new and exciting feature is the History API. Using the History API, we can fetch transactions and changes that have been made to our Swift Datastore over a particular time range. This capability allows us to update our app when we work with extensions such as widgets or sync changes to the server.

Reading the transaction history is not the only "pro" feature added to Swift Data. Let's talk about Core Data for a second.

Custom data stores in Swift Data

Core Data fundamentals included the ability to work with any data store type we wanted – XML, SQLite, CSV files, or even a remote server. Although almost all apps that implement Core Data work with SQLite as their data store, it was built to be agnostic to whatever happens underneath.

Starting with iOS 18, Apple also brings custom data stores to Swift Data.

For example, let's say that we want to base our data store on a CSV file. We start by creating a new data store configuration specifically for CSV data stores:

```swift
final class CSVStoreConfiguration: DataStoreConfiguration {
    
    typealias Store = CSVDataStore
    
    var name: String
    var schema: Schema?
    var fileURL: URL
    
    init(name: String, schema: Schema? = nil, fileURL: URL)
    {
        self.name = name
        self.schema = schema
        self.fileURL = fileURL
    }
    
    static func == (lhs: CSVStoreConfiguration, rhs:
      CSVStoreConfiguration) -> Bool {
        return lhs.name == rhs.name
    }
    
    func hash(into hasher: inout Hasher) {
        hasher.combine(name)
    }
}
```

The `CSVStoreConfiguration` class is a new data store configuration that accepts the name and the schema (similar to how Swift Data configuration setup works today), and we added an additional parameter, which is `fileURL` – the location of our CSV file.

In the `init()` function, we can also check whether the CSV file exists or whether we need to create a new one.

Notice that there's a `typealias` named `Store`, which represents a new type called `CSVDataStore`. This is the actual store class where everything happens. Let's create it now:

```swift
final class CSVDataStore: DataStore {
    
    typealias Configuration = CSVStoreConfiguration
    typealias Snapshot = DefaultSnapshot
    
    var configuration: CSVStoreConfiguration
    var name: String
```

```
  var schema: Schema
  var identifier: String


  required init(_ configuration: CSVStoreConfiguration,
    migrationPlan: (any SchemaMigrationPlan.Type)?)
    throws {
      self.configuration = configuration
      self.name = configuration.name
      self.schema = configuration.schema!
      self.identifier =
        configuration.fileURL.lastPathComponent
    }
}
```

Our `CSVDataStore` class conforms to the `DataStore` protocol and has similar properties, such as name and `schema`.

The `CSVDataStore` class must handle a persistent store's basic operations, such as inserting new items and deleting or updating existing ones.

Notice that the `init()` function includes a migration type, so we can even handle migrations when our schema changes.

To handle all of these operations, we need to implement two important methods that are part of the `DataStore` protocol – `fetch()` and `save()`:

```
func fetch<T>(_ request: DataStoreFetchRequest<T>)
  throws -> DataStoreFetchResult<T, DefaultSnapshot>
    where T : PersistentModel {

      let predicate = request.descriptor.predicate

      return DataStoreFetchResult(descriptor:
        request.descriptor, fetchedSnapshots: [],
        relatedSnapshots: [:])
          . // perform fetch operations
  }

  func save(_ request:
    DataStoreSaveChangesRequest<DefaultSnapshot>)
    throws -> DataStoreSaveChangesResult<DefaultSnapshot>
    {
      var remappedIdentifiers = [PersistentIdentifier:
        PersistentIdentifier]()
```

```
        for snapshot in request.inserted {
            // insert new items
        }
        for snapshot in request.updated {
            // update existing items
        }
        for snapshot in request.deleted {
            // delete items
        }
        return
          DataStoreSaveChangesResult<DefaultSnapshot>(for:
          self.identifier,
          remappedIdentifiers: remappedIdentifiers)
    }
```

These two functions perform all the magic underneath. In this code example, I left the function implementation empty – it is up to you to fill it in according to the specific data store implementation. Once we modify our CSV file, we can return the results to the app.

The `History` API, the `DataStore` protocol, and the ability to provide uniqueness to entities make Swift Data much more mature and capable. To get started with Swift Data, read *Chapter 2*.

Next, let's talk about an exciting improvement in SwiftUI transition.

Introducing zoom transition

This is a small improvement, but it may indicate an interesting direction Apple is taking with SwiftUI. In general, UIKit's transitioning capabilities are very robust and provide us with the flexibility to create any transition we want. Even before that, from the beginning, UIKit had some nice built-in transitions we could use to make our navigation more appealing.

In iOS 18, Apple added a new transition that allows us to navigate to a new view using a zoom animation.

Let's create an album grid that, when tapping on the album, transitions to a full album screen with a zoom animation:

```
        @Namespace() var namespace
    var body: some View {
        NavigationStack {
            ScrollView {
                LazyVGrid(columns: [
                    GridItem(.adaptive(minimum: 150)) ]) {

                    ForEach(Album.albums) { album in
                        NavigationLink {
```

```
                        Image(album.imageName)
                            .resizable()
                            .navigationTransition(.
zoom(sourceID: album.id, in:
  namespace))
                    } label: {
                        Image(album.imageName)
                            .resizable()
                            .scaledToFit()
                            .frame(minWidth: 0,
                              maxWidth: .infinity)
                            .frame(height: 150)
                            .cornerRadius(8.0)
                    }
                    .matchedTransitionSource(id:
                       album.id, in: namespace)
                }
            }
          }
        }
        .padding()
    }
}
```

This example shows a simple grid view of albums, a NavigationStack, and a NavigationLink. The idea of performing the zoom transition is to match the source (the image we tapped on) to the destination (the image we zoomed into).

We do that by adding two view modifiers:

- `navigationTransition`: We add this modifier to the source view. The source view, in our case, is the album view in the grid. We select the type of animation (currently, it's a zoom animation) and the source ID.
- `matchedTransitionSource`: We add this modifier to the destination view. In our example, the destination view is the full-screen view of the album. Again, we provide the ID of the album we want to present so SwiftUI can perform the zoom animation between these views.

Creating the match between the views allows SwiftUI to perform a nice zoom animation, similar to what we see in the Photos app. Look at *Figure 1.1*:

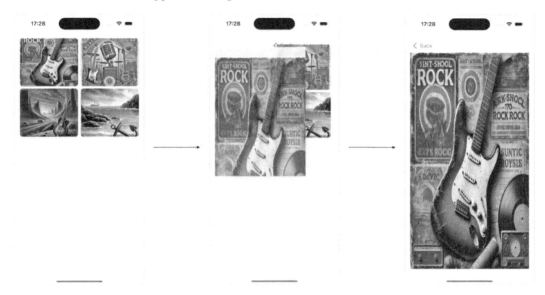

Figure 1.1: Zoom transition between photos grid and a full-screen view

Figure 1.1 shows how the zoom animation looks in a couple of frames based on the preceding code example.

Zoom transitions serve more than aesthetic purposes. They inform the user about the changes occurring on the screen, helping them stay oriented.

To read more about navigation in iOS, read *Chapter 4*.

Speaking of navigation, iPadOS navigation gained a unique and valuable capability – the floating bar.

Adding a floating tab bar

iPad is not the focus of this book. This is not because iPadOS is unimportant but because most, if not all, of the topics we discuss here are also suitable for iPadOS.

However, there are special features that are relevant to iPadOS that are worth mentioning. One of them is the float tab bar.

The tab bar has existed in iOS since its very beginning. It allows users to navigate between different sections of an app. In both iOS and iPadOS, the tab is located at the bottom of the screen. While it looks perfectly fine on small devices, a tab bar on big screens seems stretched and doesn't use the large space.

One solution for handling navigation in a iPadOS is to implement a sidebar – a view on the side that displays the different sections of the app.

In iPadOS 18, the position of the sidebar changed, and it is now located at the top of the screen, floating over the app content. Not only that; the user can also transition between a tab bar and a sidebar. Let's see how to do that in code:

```
struct ContentView: View {
    
    var body: some View {
        TabView {
            Tab("Home", systemImage: "house.fill") {  }
            
            Tab("Profile", systemImage:
                "person.crop.circle") {  }
            
            Tab("Settings", systemImage: "gear") {  }
        }
        .tint(.red)
        .tabViewStyle(.sidebarAdaptable)
    }
}
```

This code example looks straightforward but includes a view modifier called `tabViewStyle`. Currently, it has only one option to choose from – `sidebarAdaptable`. When we add this view modifier, a button is added to the tab bar that allows the user to change the layout. Let's see how it looks (*Figure 1.2*):

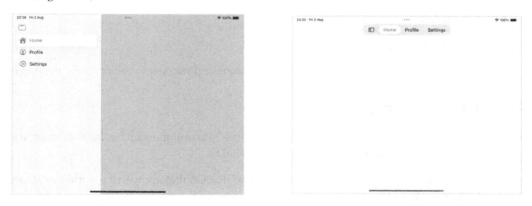

Figure 1.2: The Tab bar adapts a sidebar layout

Figure 1.2 shows the two layouts for our tab bar. The new sidebar improves the user experience and makes navigating and focusing on content easier. It also resembles Apple's apps, such as the TV app, which aligns with what users can expect from our app.

Another important aspect of SwiftUI that required improvement is scroll views. Let's go over major changes in that area.

Having more control over scroll views

Controlling and observing scroll view behavior was part of the reason why UIKit developers hadn't moved to SwiftUI yet.

Scroll views are crucial in mobile apps, not just because of the small screen, which often requires the user to scroll for more content, but also because they help reuse visible content to minimize memory usage or adjust our UI based on scroll position.

However, why is handling scroll views in SwiftUI more complex than in UIKit? We can think of two reasons:

1. **SwiftUI is relatively new**: SwiftUI is still considered to be a new framework. Think how much time it took for UIKit to become a mature framework. Obviously, we can achieve this in several years and 17 years of development.
2. **Flexibility**: Due to the imperative approach, UIKit gives us direct control over views. This means that we can adjust particular view parameters based on the scroll state. SwiftUI's declarative nature sometimes makes achieving the same level of control challenging – we don't have direct access to views. We can adjust their state using a `@State` variable or a view modifier.

These reasons lead to many workarounds that developers use to achieve the desired user experience. Fortunately, iOS 18 gives us two view modifiers that make SwiftUI scroll views more appealing than ever. We'll start with `onScrollGeometryChange`.

Observing the scroll view position

Up until now, SwiftUI hasn't provided any direct API to observe the scroll view position. Many developers had to find a workaround or use UIKit as a solution. Now, we have an `onScrollGeometryChange` view modifier that allows us to observe any change in the scroll position.

Let's say we have a `VStack` view within a scroll view, and we wish to show a **Scroll to the top** button whenever the user scrolls down to allow them to return to the top of the list.

Let's look at the following code:

```
ScrollViewReader { proxy in
    ScrollView {
        VStack(alignment: .leading, spacing: 16) {
```

```
                    ForEach(albums) { album in
                        ExtractedView(album: album)
                            .id(album.id)
                    }
                }
            }
            .overlay(alignment: .bottom) {
                if showScrollToTop {
                    Button("Scroll to top") {
                        proxy.scrollTo(albums[0].id,
                            anchor: .top)
                    }
                    .buttonStyle(.borderedProminent)
                }
            }
            .onScrollGeometryChange(for: Bool.self) {
            geometry in
                geometry.contentOffset.y <
                geometry.contentInsets.bottom + 300

            } action: { oldValue, newValue in
                withAnimation {
                    showScrollToTop = !newValue
                }
            }
        }
```

In this code example, we can see a `VStack` view inside a scroll view. The `VStack` view contains a list of albums. Notice that we have an `onScrollGeometryChange` view modifier for the scroll view itself. The view modifier has a closure that runs each time the scroll position changes with a `geometry` parameter. Within the closure, we inspect the scroll view content offset, and if it reaches a specific threshold, we show/hide the **Scroll to top** button using a specific state variable.

The `ScrollViewReader` view, which wraps the scroll view, provides a proxy to the scroll view so we can scroll to the top when the user presses the button.

We can use the `onScrollGeometryChange` method for more use cases than just toggling a button. For example, we can use it to perform a network request in an infinity list where we need to load more content from the server when the user reaches the bottom. Additional examples would be having a sticky header or a progress indicator, or even just sending analytics. These use cases were complex to implement before iOS 18 and are now extremely simple.

The improvement in the second scroll view seems to belong to the same family. Let's review it now.

Observing items' visibility

Checking whether a view is visible inside a scroll view is not easy. Up until now, we had to calculate the view frame versus the scroll view content offset, not to mention observe that during a scroll view. Lucky for us, we now have a new modifier called `onScrollVisibilityChange`.

Suppose we want to change a view while it enters our scroll view. For example, we might want to report analytics or print to the console.

Let's look at the following example:

```
ForEach(albums) { album in
    ExtractedView(album: album)
        .id(album.id)
        .onScrollVisibilityChange(threshold: 0.9) {
          visible in
            if visible {
                print("\(album.title) appears")
            }
        }
}
```

This code example shows the same album row we created in the previous example (in the *Observing the scroll view position* section). This time, we added a new view modifier to the view itself – `onScrollVisibilityChange`. This view modifier has two parameters – `threshold` and `closure` with a `Bool` parameter (named `visible` in our case). Let's review them now:

- `threshold`: The `threshold` parameter defines how much the change must occur for the closure to run. For example, a threshold of 0.2 means that we need 20% of the view to be visible or hidden before it runs the closure and reports the change.
- `closure`: The closure with the `Bool` parameter runs each time the view reaches the threshold. The `Bool` parameter contains the change – `true` for visible and `false` for hidden.

In our code example, we set the threshold to `0.9`. This means that we need to view it to reveal 90% of its size before the closure runs. Inside the closure, we check whether the view is visible before we report it to the console.

We can use this view modifier for many purposes. For example, we can perform a specific animation when the view enters, load additional information, or adjust the screen interface if needed. Something that was complex to do before is now simple to accomplish using one view modifier.

Scroll view is not the only topic we have more control of. Let's talk about texts.

Changing the text rendering behavior

Handling texts on screen was also a very mature area where UIKit provided great frameworks such as TextKit. We could manipulate texts and create almost any effect that we wanted.

In iOS 18, Apple introduced TextRenderer, a protocol that can help us change the default behavior of our texts in SwiftUI.

Let's say that we want a title with a different opacity for each line and even rotate the lines a bit. This creates a nice effect for the titles in our app. So, let's see how to do that in SwiftUI:

```swift
struct CustomTextRenderer: TextRenderer {
    
    func draw(layout: Text.Layout, in ctx: inout
      GraphicsContext) {
        for (index, line) in layout.enumerated() {
            ctx.opacity = Double(index + 1) * 0.1
            ctx.rotate(by: Angle(degrees: Double(index) *
              1))
            ctx.draw(line)
        }
    }
}

struct ContentView: View {
    var body: some View {
        Text("Great new features come to texts in SwiftUI")
            .font(.system(size: 60))
            .textRenderer(CustomTextRenderer())
    }
}
```

This code example has a new structure called `CustomTextRender`, which conforms to the `TextRenderer` protocol. We have one important function to implement – the `draw()` function. In this function, we receive an important parameter – `ctx` – the graphic context. The `TextRenderer` protocol also provides us access to the different lines and slices we have in our text. In our example, we can iterate the different lines using the `layout` parameter, change their opacity, and even rotate them.

Once we have the `CustomTextRender` structure, we can add it to our `Text` component using the `textRenderer` view modifier.

Let's see how it looks (*Figure 1.3*):

Figure 1.3: The Text component with custom text rendering

Figure 1.3 shows our text with a different opacity and rotation for each line. Adding effects to text can give a dynamic visualization for titles and paragraphs and add more life to our apps.

Next, let's see how SwiftUI has become more mature and capable than ever with positioning sub -iews from other views.

Positioning sub-views from another view

What does it mean to position sub-views from another view? While this description sounds weird and unclear, it is a nice addition to SwiftUI that can help us provide more dynamic and reusable content.

To understand what it means, let's take the following code as an example:

```
struct NewsView: View {
    var body: some View {
        Text("Major Breakthrough in Renewable Energy: New Solar Panel Technology Promises 30% Efficiency Increase")
        Text("lobal Markets React to Sudden Interest Rate Hike: Stocks Tumble Across the Board")
        Text("Historic Peace Agreement Reached: Leaders Sign Pact to
```

```
End Decades-Long Conflict")
        Text("Innovative AI Tool Revolutionizes Healthcare: Doctors
Embrace Machine Learning for Diagnosis")
        Text("Natural Disaster Strikes: Massive Earthquake Hits
Coastal City, Rescue Efforts Underway")
    }
}
```

This code example shows a view called `NewsView` with a list of `Text` components, each containing a news headline. If we look closely, we can see that there's no layout – no VStack, group, or List. We are not used to this in SwiftUI, and that's okay because that view is for display.

The `NewsView` job is to be a container for components. Let's see how we can use this container:

```
struct ContentView: View {
    var body: some View {
        
        ScrollView {
            VStack {
                Text("Latest headlines")
                    .font(.title)
                
                Group(subviews: NewsView()) { collection in
                    if let firstHeadline = collection.first
                    {
                        firstHeadline
                            .font(.title2)
                        Spacer()
                    }
                    ForEach(collection.dropFirst()) {
                      newsItem in
                        newsItem
                            .font(.headline)
                        Spacer()
                    }
                }
            }
            .padding()
        }
    }
}
```

In this example, we added a SwiftUI group, but this time, from the `NewsView` view:

```
Group(subviews: NewsView()) { collection in
```

This line creates a group that iterates over the specific view's sub-views and allows us to position and modify them ourselves.

In our example, we change the font of the first sub-view and present all the views with a spacer between them.

The ability to reposition views within other views unlocks new possibilities. For instance, we can reuse the same views but with different layouts, sequences, or styles. Treating our views as containers for smaller components makes our code more reusable.

Now, let's move to our chapter's last section – the AI revolution.

Entering the AI revolution

AI and machine learning are not new areas for Apple and the iOS platform. Apple uses AI to adjust photos taken, suggest apps to users according to their usage, optimize battery charging, and many more.

For developers, Apple provides the CoreML framework and tools such as Create ML to help users train and create their own machine learning models.

However, the rising popularity of services such as ChatGPT and Gemini proved that CoreML is insufficient, and that Apple needs to integrate AI deeper into the system.

So, what did Apple prepare for us, the developers, regarding AI in iOS 18?

Apple integrated AI into iOS 18 by letting iOS understand what's happening in the system and helping the user perform common tasks using natural language understanding, similar to ChatGPT.

For example, let's say we're working on a word-processing app and created an App Intent that allows the user to add an image to a document.

Until iOS 18, we would have had to define a specific phrase for the user to use with Siri. However, in iOS 18, the user can say something such as "Add this image to the page I'm working on," and Siri uses a set of machine-learning models to convert this phrase to our app intent model. Not only that, but Siri can also understand the current context on screen and even search our app by indexing our app content in the spotlight.

Integrating our app into Siri requires little effort. We mainly need to focus on structuring our main actions and entities. Apple Intelligence does all the rest.

To read more about using App Intents with Siri, go to *Chapter 13*.

Summary

There's no other way of looking at iOS 18 than as an exciting one. The addition of Apple Intelligence is only part of the story – Apple took care of many system and SDK aspects such as testing, persistent data, UI, and more.

In this chapter, we explored the basics of the new Swift Testing framework, learned about Swift Data improvements, and discussed enhancements in SwiftUI such as zoom transition, floating tab bar, scroll views, and text rendering. We even scratched the surface of Apple Intelligence and tried to understand how it is integrated with App Intents. By now, you should be familiar with the most exciting and new topics in iOS 18.

A few code examples are just not enough. We are developers, and we need more! So, let's jump straight into SwiftData and explore Apple's new persistent data framework in the next chapter.

2
Simplifying Our Entities with SwiftData

Let's start our journey to mastering iOS 18 with one of the most important and useful frameworks Apple has released in the last few years – SwiftData.

SwiftData is an excellent example of Swift macro usage, taking the old and beloved Core Data framework to a whole new level of simplicity and adapting it to the modern world of Swift and declarative programming.

In this chapter, we will do the following:

- Understand the SwiftData background
- Define a data model, including its relationships and attributes
- Learn about the SwiftData container and configurations
- Fetch and manipulate data using the model context
- Migrate our data to new version schemas

It's going to be a long ride with an exciting new framework! So, after the technical requirements, let's start with some background on the framework.

Technical requirements

This chapter includes many code examples, some of which can be found in the following GitHub repository:

https://github.com/PacktPublishing/Mastering-iOS-18-Development/tree/main/Chapter%202

To run them, we will need Xcode 16 or newer.

Understanding SwiftData's background

To understand SwiftData's background and its roots, it's important to go one step backward and learn about the **Core Data** framework.

Core Data has been the primary data framework for Apple platforms for many years, even before iOS was born.

Core Data was added to iOS in iOS 3, bringing the power of handling a data graph to mobile devices flexibly and efficiently. Note that I haven't mentioned the word *database* or *persistency*, and that's for a good reason. We should remember that Core Data is not an **SQLite** wrapper, even though its persistent store is based on SQLite in most cases. The primary goal of Core Data is to handle our app's data layer.

But what does it mean to handle the app's data layer? Well, most apps work with several layers – the UI, business logic, and data layer. The data layer is built upon data entities that define the core items that our app works with. For example, a to-do app can have entities such as a *list, task,* or *reminder*. A music app can have entities such as an *album, song,* or *playlist*.

The data layer defines the different entities and how they are related. For example, an album can contain many songs, and a list can contain many tasks. if there's a need for persistence, the data layer also handles how the different entities' data is saved to disk. According to our understanding of a data layer, Core Data fulfills its role as an app data layer by defining its data model, handling persistency, migrations, and even undo operations. So, if Core Data is such an excellent framework for handling data, why do we need SwiftData?

Core Data is a great framework, but it was designed for different times when we used to code with Objective-C, and UIkit hadn't even been created. Ever since then, the iOS development world has changed significantly – we now have Swift, and moreover, we have SwiftUI. Even though Core Data has received updates to support Swift and SwiftUI, it still felt outdated in a world of type-safety, multithreading, and declarative programming. Fetching and observing data changes have become cumbersome in Core Data, as we use design patterns more suitable for the UIkit/Objective-C era. In this context, SwiftData promises to bring a modern, straightforward framework to handle data much more flawlessly, using the full power of Swift and Combine.

One of the best things about SwiftData is that it uses *Swift macros* – the same Swift macros we learned about in *Chapter 10*. The macros help us elegantly implement SwiftData without using boilerplate code.

It's time to get into business and create our first SwiftData models!

Defining a SwiftData model

Usually, when discussing a data framework, it is common to start with the basic setup. However, this time, we will begin with the model itself. Why is that? Because I want to demonstrate how simple and easy it is to convert an existing data model to a SwiftData model, using the following piece of code:

```
import SwiftData

@Model
class Book {
    var author: String
    var title: String
    var publishedDate: Date

    init(author: String, title: String, publishedDate:
      Date) {
        self.author = author
        self.title = title
        self.publishedDate = publishedDate
    }
}
```

In this code, we see a standard `Book` class, with the addition of a macro named `@Model`. Before we expand the `@Model` macro and see what it does precisely, let's focus on what happens when we add it.

Adding the `@Model` macro is all it takes to convert a regular class into a model backed with a persistent store. Similar to how Core Data entities work, the class name is the entity name, and its variables are the entity attributes.

When we compare that to Core Data, we can see that the model declaration process is backward – in Core Data, we declare the model in the model editor and then generate its class, whereas in SwiftData, we take a regular class and make it a model.

But what does `@Model` macro really do? Let's expand it and see.

Expanding the @Model macro

We already know what a Swift Macro is capable of, and SwiftData is a great chance to explore a new macro implementation.

To expand the macro, we can right-click on the `@Model` name and select **Expand Macro** from the pop-up menu. The class body now looks like this:

```
@Transient
private var _$backingData: any SwiftData.BackingData<Book>
  = Book.createBackingData()

public var persistentBackingData: any
  SwiftData.BackingData<Book> {
    get {
        _$backingData
    }
    set {
        _$backingData = newValue
    }
}

static var schemaMetadata:
  [SwiftData.Schema.PropertyMetadata] {
  return [
    SwiftData.Schema.PropertyMetadata(name: "author",
      keypath: \Book.author, defaultValue: nil, metadata:
      nil),
    SwiftData.Schema.PropertyMetadata(name: "title",
      keypath: \Book.title, defaultValue: nil, metadata:
      nil),
    SwiftData.Schema.PropertyMetadata(name:
      "publishedDate", keypath: \Book.publishedDate,
      defaultValue: nil, metadata: nil)
  ]
}

required init(backingData: any SwiftData.BackingData<Book>) {
  _author = _SwiftDataNoType()
  _title = _SwiftDataNoType()
  _publishedDate = _SwiftDataNoType()
  self.persistentBackingData = backingData
}

@Transient
private let _$observationRegistrar = Observation.
ObservationRegistrar()

struct _SwiftDataNoType {
```

```
}

extension Book: SwiftData.PersistentModel {
}

extension Book: Observation.Observable {
}
```

So, what happened to our beautiful and minimal `Book` class? It appears that the `@Model` macro has been quite active here.

To simplify it, let's try to break it down:

- **Our `Book` class now conforms to two protocols**: These fundamental protocols are `PersistentModel` and `Observable`. The `PersistentModel` protocol helps SwiftData work with our style and access its attributes. The `Observable` protocol allows us to be notified of changes to the data.

- **Having backing data and metadata properties**: If we go even deeper with our exploration and try to understand what the `PersistentModel` protocol is, we will discover that it requires the implementation of two variables –`backingData` and `schemaMetaData`. We can see their implementation directly in our macro-expanded code. These variables help SwiftData to store and retrieve our entity information specifically for our properties. And perhaps this is where the real power of Swift Macro comes into play – the ability to generate code that is custom-made for our class.

- **We have property macros**: If we look at the class properties, we can see that they have their macros now. Expanding them reveals that they have now become a computed variable, so we can store and retrieve data not from just our memory but also from our backing data store:

    ```
    @_PersistedProperty
    var author: String
    @_PersistedProperty
    var title: String
    @_PersistedProperty
    var publishedDate: Date
    ```

Additional lines of code wrap everything together, such as the observation and registering attributes.

Is this complicated? A little bit. But that's one of the benefits of having a macro – to simplify complex implementations. What's important to understand is that every class marked with a `@Model` macro immediately receives a store of its own and is added to the SwiftData schema.

However, to add a more complex data model, we need to be able to define relationships between our models. Let's see how it works.

Adding relationships

Unlike real life, in SwiftData, relationships are simple.

A **relationship** is a database scheme that defines how entities are linked to each other, and in Core Data, we have two types of relationships – **to-one** and **to-many**. In short, a *to-one* relationship means that we will have one instance of the other kind for each entity instance. An example of that would be cars and engines – every car has one, and only one, engine, so that will make the relationship between them a *to-one* relationship. However, cars and wheels have a *to-many* relationship because a car can have multiple wheels.

Even though the explanation is simple enough, it gets even simpler in SwiftData. If we want to define a relationship between models, we just need to create another variable, as shown here:

```
@Model
class Book {
    var title: String
    var publishedDate: Date
    var author: Author
    var pages: [Page]

    init(author: Author, title: String, publishedDate:
      Date) {
        self.title = title
        self.publishedDate = publishedDate
        self.author = author
        self.pages = []
    }
}
```

In our example, we added the following two properties to the Book class:

- Author: This is a *to-one* relationship to the Author entity because, in our case, each book has only one author
- pages: In the case of Page entity, we have a *to-many* relationship, since a book can contain multiple pages

One thing to note is that we also need to mark both Page and Author entities with the @Model macro, as they must be part of our schema. This can be seen in the following code:

```
@Model
class Author {
    var name: String

    init() {
```

```
            self.name = ""
        }
    }

    @Model
    class Page {
        var content: String
        var order: Int

        init(content: String, order: Int) {
            self.content = content
            self.order = order
        }
    }
```

Is adding models so simple? The short answer is, yes! Linking entities to each other in SwiftData is as easy as adding a property.

The longer answer is, well, we'll have to do extra work to customize the relationships a little bit. Let's meet the @Relationship macro.

If you're familiar with Core Data relationships, you probably know there is more than declaring *to-many* and *to-one*.

> **To-many and to-one relationships**
>
> To-one relationships represent associations between entities where one instance of an entity is related to another single instance of a different entity. Conversely, to-many relationships represent associations where one instance of an entity can be related to multiple instances of another entity. For example, in a bookstore database, a to-one relationship could connect a "book" entity to an "author" entity, as each book has one author. In contrast, a to-many relationship could connect a "book" entity to a "category" entity, as a book can belong to multiple categories.

We can customize our relationship using the @Relationship macro in two primary ways.

Let's start with defining the deletion rules.

SwiftData relationship deletion rules

What happens to pages and author entities if we delete a book? Logically, all the book pages need to be deleted, but the author needs to be retained because they might be linked to another book. We can represent this logic with *deletion rules*; if you're familiar with Core Data, it is basically the same as in SwiftData.

This is how we can define the logic to a property in SwiftData:

```
@Relationship(.unique, deleteRule: .cascade) var pages: [Page]
```

In our code example, we defined the delete rule as `cascade`.

We have four different deletion rules:

- `cascade`: Deletes any related objects
- `deny`: Prevents deletion of an object if it contains one or more references to other objects
- `nullify`: Nullifies the related object's reference to the deleted object
- `noAction`: In this case, nothing will happen to the other object

We should remember that a deletion rule is not arbitrary; it should be based on our app business ideas.

For example, the reason why a book has a *to-one* connection to an author sounds logical, but there are books with co-authors as well. So, this is something that should be aligned with our product manager.

Most of us are more familiar with the term *one-to-many* than *to-many*. This is because relationships between objects go both ways – the fact that each book has one author doesn't mean that each author has only one book. So, as part of the relationship definition, we also need to define its inverse relationship.

Defining the inverse relationship

Why do we need to define the inverse relationship? We need to realize that relationships always have two sides (like in real life!), and we need to maintain them to have a proper data schema.

When establishing a relationship between a book and its pages, it's better to define the inverse relationship as well. This way, we can create a proper reference back to the book.

Let's see how to create an inverse relationship between a book and its pages through the following code:

```
@Model
class Book {
    ...
    @Relationship(inverse: \Page.book) var pages: [Page] =
        []
    ...
}

@Model
class Page {
    var content: String
    var book: Book?
```

```
    init(content: String) {
        self.content = content
    }
}
```

Looking at the code, we can see that we define the relationship as a keypath:

```
\Page.book
```

A keypath can help us avoid typos and mistakes when defining the inverse property.

Moreover, if we add a new page to the `pages` property, SwiftData will automatically set the Page's `book` property to the new book:

```
let newPage = Page(content: "Swift Data")
newPage.book = book
// book.pages property contains 'newPage'
```

SwiftData knows how to do that using our inverse declaration.

The inverse relationship may sound like an obvious feature – if we have a book with several pages, and each page is related to a book, isn't it obvious that the `book` property in the `page` class is the inverse relationship? However, in reality, it's not obvious. There are several real-world use cases when relationships can be much more complex.

Let's take, for example, the data structure of a folder tree – each folder has its sub-folders. This means that a folder has a *to-one* relationship to its parent and a *to-many* relationship to its children. Let's see that in the code:

```
@Model
class Folder {
    
    var parent: Folder?
    @Relationship(inverse: \Folder.parent) var subFolders:
      [Folder]
    var name: String
    var id: UUID
    
    init(parent: Folder? = nil, subFolders: [Folder], name:
      String, id: UUID) {
        self.parent = parent
        self.subFolders = subFolders
        self.name = name
        self.id = id
    }
}
```

This example demonstrates what a `Folder` class looks like when trying to create a multi-level hierarchical structure. In this case, we must define the inverse relationship to avoid cycles.

Now that we know how relationships work in SwiftData, let's see more ways to customize our model, using the `@Attribute` macro.

Adding the @Attribute macro

So far, we have learned how to declare new entities, properties, and even relationships between our entities. It looks like we can do anything with our data entities! Now, it's essential to drill down to the property level.

Along with our `@Model` and `@Relationship` macros, we now have the `@Attribute` macro to define the behavior of a specific property.

If you remember from Core Data, each attribute has an inspector window where we can configure an attribute's behavior (*Figure 2.1*):

Figure 2.1: The Attribute inspector in Core Data

Figure 2.1 shows what it looks like when we select one of the attributes (`firstName` in this example) and how we can customize its behavior.

We can define some of these settings in SwiftData as part of the property declaration. For example, the Optional feature, as seen in *Figure 2.1*, is defined by marking a property as Swift optional type, and the default value is part of variable initialization:

```
var firstName: String? = "MyName"
```

However, other settings need to be declared as part of the `@Attribute` macro.

Let's start with the most common one, `unique`, and making attributes unique is an important feature of many databases, including *SQLite*.

The following are a few reasons why:

- **Setting up a primary key**: A primary key represents a record's unique identifier. We use a primary key to ensure that there are no duplicates in our table.
- **Supporting indexing**: Unique attributes can help us index our database for searching and retrieval.
- **Helping with data validation**: Utilizing unique attributes goes beyond primary keys and extends to other distinctive attributes, enhancing our ability to validate data during insertion.

Even though *SQLite* supports unique attributes, Core Data doesn't have a built-in way to support unique identifiers, derived from its design philosophy to offer complete flexibility to developers.

Conversely, SwiftData supports unique attributes out of the box:

```
@Attribute(.unique) var id: UUID = UUID()
```

Adding the `@Attribute` macro with the `.unique` option makes our database's specific property values unique.

`UUID` is a classic example of a unique value for a property, but we can apply that to any other type of property, such as user IDs and names.

But what does it really mean to make a property *unique*? What will happen when we try to insert an instance with an already existing unique attribute?

In the case of a unique property, SwiftData performs an **upsert**, which is also called an `INSERT` or `UPDATE` operation. This means that if an instance with a unique value already exists, SwiftData will not create a new object in its store but, rather, update the existing instance.

Declaring a property as unique using the `@Attribute` macro is straightforward. However, sometimes we need something more sophisticated. For example, let's say we have a `Book` model with `name` and `publicationName` properties. In our case, we can have two books with the same name or the same `publicationName`, but we can't have two books when both properties are identical. The combination of `publicationName` and `name` creates the book's unique identity.

One solution is maintaining a property that tries to build a unique ID from these two properties. Another elegant option is to use the `#Unique` macro to define more complex uniqueness requirements:

```
@Model
class Book {

    #Unique<Book>([\.name, \.publicationName])

    var publicationName: String = "Packt"
    var name: String
}
```

In this code example, we enforce the uniqueness of the `Book` model by combining two key paths. Just like the attribute parameter, `.unique`, if we try to insert a new book instance when we already have one with the same name and publication name, SwiftData will perform an `upsert` operation and update that instance.

Even though SwiftData handles unique attributes well, it is important to ensure we carefully pick unique attributes and key paths according to an application's requirements. Too many unique attributes can cause complexity and performance issues.

Unique attributes are great for simplifying the task of handling duplicate instances. Another attribute feature that can make our life simpler is *transient*.

Going non-persistent with transient

The nice thing about working with SwiftData is that all properties automatically become the entity's attributes and are saved persistently to the local data store. However, sometimes, there are cases where we want to have a property that is *memory-only* and not saved persistently. A formatted version of one of the values is an excellent example of such a property. One option to achieve that is to create a function or a computed variable, and then return a value based on the relevant property. However, there are other cases where a computed variable or a function is not a convenient solution. Let's say we want a temporary counter or to maintain a flag relevant only to an application's current life cycle.

For these kinds of cases, we have a *transient* attribute. Transient attribute is not a new idea – Core Data had transient properties from its early versions. Since SwiftData is based on Core Data fundamentals, it supports transient properties out of the box.

Here is how we declare a *transient* property in SwiftData:

```
@Transient
var openCounter: Int = 0
```

In this code snippet, the `openCounter` variable will not be saved locally to the persistent store and will be initialized each time we fetch the entity from our database.

Transient properties may sound like a minor feature, but there are many cases where it really makes the difference, and the transient macro provides this flexibility. Full names or calculated ages are great examples of that.

Exploring the container

Until now, we discussed how to declare the different entities using the `@Model` macro, define their relationships using the `@Relationship` macro, and customize their attributes using the `@Attribute` macro.

However, we haven't discussed setting up SwiftData to work with a schema and a persistent store.

When we delved into the `@Model` macro, drawing parallels with Core Data was straightforward and remains so now. In Core Data, we set up the stack using `NSPersistentContainer`, which encapsulates the different components, such as the data model, the store, and the context, into one stack that we can work with.

In SwiftData, we use `ModelContainer`, which has the same responsibility.

Let's try to understand how it works.

Setting up ModelContainer

`ModelContainer` is essential for working with SwiftData. The reason is that SwiftData has three main components that the container encapsulates and wraps together:

- **The data model**: This is what we defined in the *Defining a SwiftData model* section, adding the `@Model` macro to our entities
- **The store**: The backend store where we will save our data
- **The context**: This is our link to the store and the sandbox, where we can add, edit, and delete different records

Here is the basic and minimal way to create a container:

```
var container: ModelContainer = {
    do {
        return try ModelContainer(for:
            Schema([Book.self, Author.self, Page.self]) )
    } catch {
        fatalError("Could not create ModelContainer:
            \(error)")
    }
}()
```

In this code, we create an object from the `ModelContainer` type and provide it with the three models we made earlier in the *Defining a SwiftData Model* section.

Note that, in our case, we have one parameter, `Schema`, which holds all the different models relevant to our container – `Book`, `Author`, and `Page`.

The fact that we need to provide a list of models may raise some eyebrows – why do we need to do that? Can't Xcode locate all the models and add them automatically? The `@Model` macro indeed expands code at compile time, but it doesn't mean SwiftData is aware of all our entities when we set it up at the beginning of the app run. So, whenever we add a new model, we must add it to the list of models in our `Schema` parameter.

Regarding including the `Book` entity independently from the `Author` entity – when we add the `Book` model to the models' list, it automatically consists of all associated models, including those related to the models further down the hierarchy. It means that, theoretically, we can include only the root object while doing something like that:

```
Schema([Book.self])
```

And that will be enough to include `Author` and `Page`.

So, what will we do with the container instance we just created? Let's see in the next section.

Connecting the container using the modelContainer modifier

Now that we have a model container, we want to link it somehow to our UI so that we can start using it.

To do that, we will use the `modelContainer` modifier to connect the container to our scene:

```
var body: some Scene {
    WindowGroup {
        ContentView()
    }
    .modelContainer(container)
}
```

In our code example, we add the `modelContainer` modifier to our `WindowGroup`, thus making it available to the whole app.

Instead of creating a connector and connecting it to `WindowGroup`, we can use another `modeContainer init` method and pass only the list of entities:

```
.modelContainer(for: [Book.self, Author.self, Page.self])
```

Passing the list of entities can be a simple and easy way of setting up a container. So, why do we need the `ModelContainer` class? The simple answer is, as always, to provide more customization. Let's see how!

Working with ModelConfiguration

The `ModelContainer` offers us more than just scheme passing; it empowers us to configure our `SwiftData` store for specific models and customize it to our particular requirements.

To do that, we will use the `ModelConfiguration` struct, as follows:

```
var modelContainer: ModelContainer = {
      do {
          let schema = Schema([Book.self, Author.self,
            Page.self])
          let modelConfiguration =
            ModelConfiguration(schema: schema,
              isStoredInMemoryOnly: true)
          return try ModelContainer(for: schema,
            configurations: [modelConfiguration])
      } catch {
          fatalError("Could not create ModelContainer:
            \(error)")
      }
}()
```

Let's try to understand what is happening in this code snippet. First, we create a schema with a list of our models. Then, we declare a model configuration struct, pass the schema, and set its backend store to in-memory. Finally, we return a model container based on our schema and a set containing the configuration we just created.

All this process feels a little bit awkward, clumsy, and redundant – why do we need to create a configuration if we're passing the same schema again? And why is it a set? The main configuration idea is to provide different behavior for a different set of models.

Here's an example. Imagine we have a brainstorm sketch app. We want to sketch and store our concepts in the app's persistent storage while all drawings on the whiteboard canvas remain in memory.

In this case, we can create two configurations, one for in-memory and one for persistent storage and **CloudKit** integration:

```
var modelContainer: ModelContainer = {
      do {
          let brainstormDataConfiguration =
            ModelConfiguration("brainstorm_configuration",
              schema: schemaForBrainstorm,
              isStoredInMemoryOnly: true)
          let projectsDataConfiguration =
            ModelConfiguration("projects_configuration",
              schema: schemaForProjects,
```

```
                cloudKitDatabase: .automatic)
            return try ModelContainer(for: fullSchema,
                configurations: [brainstormDataConfiguration,
                projectsDataConfiguration])
        } catch {
            fatalError("Could not create ModelContainer:
               \(error)")
        }
    }()
```

In our example, we created two different schemas – a list of models for brainstorming and a list of models for the user projects.

Based on these models, we create two different configurations. The idea brainstorming configuration is saved in memory, and the projects' configuration is saved locally and syncs to CloudKit.

Working with two different configurations and schemas for two app features is a great example of model configuration usage. We can use the model configuration for additional customization, such as the following:

- Different store files
- Different group containers
- Different auto-saving mechanisms

However, let's suppose we don't need the model configuration to configure different behavior for different groups of models. In that case, we can work directly with the model container and initiate it with the entire schema.

We now know how to declare and group our models for a schema to be used in a model container. But there's one crucial thing missing – how to insert, update, and fetch data. We will do that by placing the missing piece in the puzzle – the context.

Fetching and manipulating our data using model context

Developers familiar with Core Data are also familiar with the idea of **context**. Context is our data sandbox. This is the place where we can manipulate and fetch data, and it's the link between our models and the persistent store.

To gain access to our context from our SwiftUI view, we can use an environment variable named `modelContext`:

```
struct ContentView: View {
    @Environment(\.modelContext) private var modelContext
}
```

The `modelContext` environment variable is available whenever we set up our scene using the `modelContainer` modifier.

In non-SwiftUI instances, we can access the context using our model container `mainContext` property:

```
let modelContext = modelContainer.mainContext
```

To understand how to work with a model context, we'll start with the most basic operation, saving new objects for our store.

Saving new objects

At the beginning of this chapter, in the *Defining a SwiftData model* section, we learned that our models are just Swift classes marked with a `@Model` macro.

The way we define a model in SwiftData also means that the creation of a new instance is straightforward for us:

```
let newBook = Book(name: "Mastering iOS 18 - the future")
```

Our next step is adding that book instance to our context:

```
modelContext.insert(newBook)
```

Adding `newBook` to the model context doesn't necessarily mean it is being saved to our persistent store, but it does mean that it is in our context and is ready to be pushed forward to our store. Having our entity in our context helps us manage the interaction between our app and the underlying persistent store. In our context, we can make changes, adding and deleting information without actually saving these actions to our data store. The context is beneficial when working with concurrency operations or when we want to manage undo operations.

To actually save to the persistent store, we can use the context `save()` method:

```
try? modelContext.save()
```

The `save()` method pushes changes to the store for each model, according to its configuration.

The way the `save()` method works resembles how Core Data works. But there's one difference here. SwiftData allows us to have an *auto-save* feature for the model container:

```
var body: some Scene {
        WindowGroup {
            ContentView()
        }
        .modelContainer(for: Book.self, isAutosaveEnabled:
            false)
    }
```

In our code example, we set the `isAutosaveEnabled` parameter to `false`. By default, SwiftData auto-saves every change we make to the persistent store, so there's no need to call the `save()` function unless you have a perfect reason.

Due to performance considerations, SwiftData doesn't save every single time we perform a change to the context but, rather, in the following two situations:

- During the app life cycle – for example, when moving from the foreground to the background
- In a certain time period after we perform the change

Now that we know how to create and insert new objects, we can move on to fetching.

Fetching objects

Fetching objects in SwiftData is slightly different than what we know from Core Data, as there are two primary ways to retrieve data.

The first way is to fetch an object, or objects, *based on a predicate* as part of an app flow – for example, fetching objects to sync with the server or to make some kind of calculation.

The second way is to fetch objects *based on a query* and bind them to the SwiftUI view. An example would be when we want to bind a collection of objects to a list.

Let's go over both ways and explore new structures and macros that SwiftData brings to our project.

Fetching objects using FetchDescriptor

`FetchDescriptor` is a struct equivalent to `NSFetchRequest` in Core Data.

Like `NSFetchRequest`, `FetchDescriptor` also works with a specific type of object; to use it, we can pass an optional predicate and sort descriptor.

Here's an example of how to use `FetchDescriptor`:

```
let fetchDesciprtor = FetchDescriptor<Book>(predicate:
  #Predicate { $0.name == "My Book"})
       let book = try?
         modelContext.fetch(fetchDesciprtor).first
```

If you look closely, you can see that `FetchDescriptor` is not the only new type we encounter in this context, as we also have a new `Predicate` macro that creates `PredicateExpression` (a new type in iOS 17).

Unlike the familiar `NSPredicate`, the `Predicate` macro works a little bit differently. Instead of creating a query, we have a closure where we define the condition of the return instances, like the array filter method.

The following example returns books with more than 10 pages:

```
let fetchDesciprtor = FetchDescriptor<Book>(predicate:
   #Predicate { book in
         return book.pages.count > 10
      })
```

Using the `#Predicate` macro is simple and doesn't require us to use a special syntax to perform complex queries.

In most cases, we won't have to use `FetchDescriptor`. If we want to connect data to our SwiftUI views, SwiftData has a better solution – the `@Query` macro.

Connecting data to a view using the `@Query` macro

Data is there to be seen. Showing information to the user is perhaps the most common task for iOS developers, and SwiftData's goal is just to simplify that.

As part of the SwiftData package, we now have an `@Query` macro that helps us present information in SwiftUI views:

```
@Query private var books: [Book]

var body: some View {
List {
      ForEach(books) { book in
         Text(book.name)
      }
   }
}
```

This example displays a simple list of `Book` items based on the `books` variable. The `@Query` macro before the variable declaration makes the variable a state of the view, ensuring that data is constantly updated. This means that we get an instant UI update whenever we insert a new book into our persistent store.

This is pretty remarkable for just one additional word!

The `@Query` macro also has two important additional features – filter and sorting.

Filtering the query

The chances that we will fetch *all* the items of a particular entity are pretty low, and the previous example of fetching all the books and presenting them is more common in tutorials and demo presentations.

In real life, we want to filter our queries. To do that, we can use `#Predicate`, which we learned about in the *Fetching objects using FetchDescriptor* section:

```
@Query(filter: #Predicate<Book> {
        $0.pages.count > 300
   }) private var bigBooks: [Book]
```

In this example, we added a filter to our `@Query` that returns only books that contain more than 300 pages.

Of course, we can perform even more complex queries by upgrading our Swift expression inside the predicate:

```
@Query(filter: #Predicate<Book> {
        $0.pages.count > 300 && (!$0.isRead ||
        $0.isFavorite)
   }) private var bigBooks: [Book]
```

In this example, we filter books that contain more than 300 pages, but this time, we also want to receive books that we haven't read or are marked as favorites. The fact that we use a Swift expression to filter our results makes our queries more descriptive and powerful than `NSPredicate`.

However, when displaying data in a list, it is not enough to filter it; it is also crucial to sort it. That's the job of our second main `@Query` feature.

Sorting the data

Sorting is an essential aspect of presenting information to our users. We should remember that sorting is not a lightweight task; it requires a complex algorithm to be done efficiently.

That's why we need to ensure that we can sort by a property whose type conforms to the `SortComparator` protocol, introduced as part of iOS 15.

Let's see how we can sort our filtered books:

```
@Query(filter: #Predicate<Book> {
        $0.pages.count > 300
   }, sort: [SortDescriptor(\Book.name),
        SortDescriptor(\Book.pages.count)]) private var
        bigBooks: [Book]
```

In this example, we pass an array of `SortDescriptor` – we first sort by the book name and then by its number of pages. It's pretty easy to use `SortDescriptor` – we initialize it using a key path to the desired property, just like in the preceding example.

Performing sorting with SwiftData is extremely simple. However, under the hood, it requires running algorithms that must be optimized for performance in order to work efficiently. We don't need these optimizations when working with 100 or 200 records. However, there are cases when our data store contains thousands of records. For these cases, we need to index our data.

Adding the #Index macro for performance

Before we index our data, let's try to understand what it means exactly. When performing read operations such as sorting or querying, we expect our app to work seamlessly with thousands of records. Obviously, performing a full-table scan to find a book named *Mastering iOS 18* is inefficient. So, what do we do? Like a book index, the database index contains keys that help it locate a specific record. For example, if we want to index our book's name property, we can create a data structure, such as a B-tree, which can help us locate the exact instance according to its name.

In SwiftData, we don't need to create any structure to index our data. All we need to do is add the #Index macro to our model:

```
@Model
class Book {

    #Index<Book>([\.name], [\.name, \.publicationName])
    var publicationName: String = "Packt"
    var name: String
}
```

If the preceding code looks familiar, that's because we did something similar when we added the #Unique macro to our model in the *Adding the @Attribute macro* section.

In this case, we decided to add two indexes to our model:

- The first is to index the name property, allowing an app to sort records by name or query data for a specific book name.
- The second index is based on the combination of the name and publicationName properties

If you remember from the *Adding the @Attribute macro* section, we decided that this combination defines our book's uniqueness. Creating an index for this combination can help us quickly find a specific book when needed.

Indexing looks like magic – we add another key path to the list of indexes, and everything works faster. So, why not do that for all properties? What's the catch?

It's because indexing comes with a price. First, we need to duplicate some of our data. If we need to index the name property, we need to create a structure that contains all the names. This results in additional storage for our app. But adding an index doesn't stop with storage – it also affects performance.

Indexing is not a one-time operation, as it requires maintenance. Each `insert`, `update`, or `delete` operation requires SwiftData to maintain the index structure, impacting the operation performance.

In summary, indexing is a great SwiftData feature. However, use it carefully and balance its benefits with its costs.

We've learned so many things so far! We've learned how to define models, create instances, fetch them, and connect them to the UI.

But we know that maintaining a persistent store is much more than that. Our first app version is so much different than our 50th version, and it also means that our data schema will change during our app version's life cycle. But what should we do once we already have a store full of data? That's our next topic – how to perform data migration.

Migrating our data to a new schema

Data migration is not a weird expression for those who have worked with Core Data. It is obvious that we need to change our schema as our app evolves.

There are two types of migrations – *lightweight* and *custom* migration. With lightweight migration, we perform changes that don't require custom logic. For example, adding an entity, a property, and a relationship are all good examples of lightweight migration. Conversely, changing a property type, making a property unique, and creating a new property based on other properties are examples of custom migrations. Now that we know what migration types we have, it's important to understand when it is relevant to perform a migration.

When we're in our development stage, migration is unnecessary before we have an official version of our app on the App Store. We only need to perform migration when an end user holds a version with an older schema. This also means that if we perform schema changes in several versions, we must ensure that SwiftData knows how to migrate throughout all these versions.

Now, let's discuss how SwiftData migration works and what the essential migration components are.

Learning the basic migration process

A SwiftData migration has three main components:

- `VersiondSchema`: Describes a specific schema version
- `MigrationStage`: Describes the migration process between two versions of the same schema
- `SchemaMigrationPlan`: Describes how the schema migration stages are based on the migration stages

Let's try to describe how everything is connected, using *Figure 2.2*:

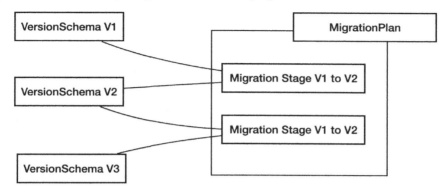

Figure 2.2: A migration process between three different versions

Figure 2.2 shows three different version schemas for three different versions. We create a migration stage each time we migrate the app from one version to another. Once we have the various stages, we can wrap them into one big migration plan.

Returning to our book's app, let's try to migrate our schema to support a `subtitle` for our Book entity.

First, we need to create our version schemas.

Creating a version schema

To migrate our book to a new schema, we need to create two version schemas – the first is our current schema, and the second is the destination schema:

```
enum BookSchemaV1: VersionedSchema {
    static var versionIdentifier: Schema.Version
    { return .init(1, 0, 0) }

    static var models: [any PersistentModel.Type] {
        [Book.self]
    }

    @Model class Book {
        var name: String

        init(name: String) {
            self.name = name
        }
    }
}
```

```
enum BookSchemaV2: VersionedSchema {
    static var versionIdentifier: Schema.Version
    {return .init(1, 1, 0) }

    static var models: [any PersistentModel.Type] {
        [Book.self]
    }

    @Model class Book {
        var subtitle: String = ""
        var name: String

        init(subtitle: String, name: String) {
            self.subtitle = subtitle
            self.name = name
        }
    }
}
```

In this code, we created two enums that conform to the `VersionedSchema` protocol. As part of the protocol definition, we need to define the version identifier and what models will change.

We added a new `subtitle` property to the second version in this case. We need to update the schema we use across the app, with the new property included.

Our next step is to define the different stages and the migration plan.

Creating the migration stages and plan

We should consider the versioned schemas as the building blocks of our migration process. *Figure 2.2* shows that we create the migration stages based on the versioned schemas.

Here's an example of a migration stage:

```
static let migrateV1toV2 =
    MigrationStage.lightweight(fromVersion:
    BookSchemaV1.self, toVersion: BookSchemaV2.self)
```

The `migrateV1toV2` stage handles the migration from `BookSchemaV1` to `BookSchemaV2`. Note that this is a lightweight migration – we only added a property, so this is all that we need to create the stage.

What about a custom migration? With a custom migration, we need to provide a closure that handles data before and after the migration stage, where we perform all the required changes.

Here's an example of a custom transition from version V2 to V3, where we have removed the subtitle property and added it as part of the book name:

```
static let migrateV2toV3 =
   MigrationStage.custom(fromVersion: BookSchemaV2.self,
   toVersion: BookSchemaV3.self, willMigrate: { context in
        if let books = try?
           context.fetch(FetchDescriptor<Book>()) {
              for book in books {
                 let newName = book.name + " " +
                    book.subtitle




book.name = newName
              }
        }
        try? context.save()
    }, didMigrate: nil)
```

As we can see in the code example, our `willMigrate` closure receives a context to work with, and SwiftData performs that closure when needed.

We fetch all the books and assemble a new name from the book name and its subtitle property. At the end of the closure code, we call `context.save()`.

Now that we have both migration steps, we can create our migration plan:

```
enum MyMigrationPlan: SchemaMigrationPlan {
    static var schemas: [VersionedSchema.Type] {
        [BookSchemaV1.self, BookSchemaV2.self,
          BookSchemaV3.self]
    }

    static var stages: [MigrationStage] {
        [migrateV1toV2, migrateV2toV3]
    }

    static let migrateV1toV2 =
      MigrationStage.lightweight(fromVersion:
      BookSchemaV1.self, toVersion: BookSchemaV2.self)

    static let migrateV2toV3 =
      MigrationStage.custom(fromVersion: BookSchemaV2.self,
      toVersion: BookSchemaV3.self, willMigrate:{context in
```

```
            if let books = try?
              context.fetch(FetchDescriptor<Book>()) {
                for book in books {
                    let newName = book.name + " " +
                      book.subtitle
                    book.name = newName
                }
            }
            try? context.save()
    }, didMigrate: nil)
}
```

The migration plan is just another Enum conforming to `SchemaMigrationPlan`, with static variables describing the list of schemas and stages (not something we haven't seen before).

Now, we have a migration plan, but SwiftData doesn't know what to do with it. Our next step will be connecting the migration plan to our SwiftData container.

Connecting the migration plan to our container

Connecting the migration plan to our container is perhaps the most straightforward step in the process.

The `ModelContainer` struct has a `migrationPlan` property specifically for that, and we need to pass the migration plan Enum type:

```
return try ModelContainer(for: schema, migrationPlan:
  MyMigrationPlan.self, configurations:
    [modelConfiguration])
```

Note the way that migrations work in SwiftData in terms of a language paradigm. We don't have to initialize anything, since we only pass the schemas, stages, and plan types. The reason is the way SwiftUI works – since we work in an immutable environment, it is much more convenient to work with static variables and types instead of instances.

Migration in SwiftData is not a simple task. It involves conforming to multiple protocols, maintaining schema versions, and understanding how a store is built to switch between lightweight and custom migration.

But this is because migration, in general, is a complex and sensitive process. Trying to carefully plan beforehand how our schema looks can reduce the number of schema versions and stages, easing our process when considering that we will have to migrate our store at some point.

Summary

SwiftData holds significance for iOS developers looking to support iOS 17 and above, representing a natural progression from Apple's previous framework, Core Data. Within the context of a declarative Swift environment, SwiftData aligns more seamlessly than before.

In this chapter, we've learned about SwiftData's background, defined the different SwiftData models, created relationships, and customized the model attributes. We moved on to the container – a component that wraps everything together, performs fetches, and saves. Lastly, we migrated our data from different schema versions using lightweight and custom migrations. Throughout the chapter, we saw the heavy use of Swift macros and protocols, which are more suitable for the modern world of Swift compared to Objective-C.

That's a lot for one chapter! Remember that the data layer is complex to manage and maintain, and there's much more to learn. The data layer is one side of our project; the other side is, of course, the UI. To complete our understanding of the data layer, it's essential to explore how the UI can monitor changes. This is why our forthcoming chapter will focus on the observation framework.

3
Understanding SwiftUI Observation

In *Chapter 2*, we discussed SwiftData, an essential framework for our data management. However, for data management to be effective, we need something on the other side that can observe changes and display them for the user.

SwiftUI contains tools that allow us to observe these changes effectively and bind them to actions and UI updates. However, these tools have become complex and confusing over the years.

Now, we're about to explore how observation has become significantly more straightforward, all while delving into the heart of SwiftUI's data flow.

In this chapter, we will cover the following topics:

- Go over the SwiftUI observation system and discuss its problems
- Add the `@Observable` macro and learn how it works
- Discuss observing properties, including computed variables
- Work with environment variables and adapt them to the new framework
- Talk about the new `@Bindable` property wrapper
- Learn how to migrate our app to work with the Observation framework

Are you ready to start?

Technical requirements

This chapter includes many code examples, some of which can be found in the following GitHub repository: `https://github.com/PacktPublishing/Mastering-iOS-18-Development/tree/main/Chapter3`

To run them, we will need Xcode 15 or newer.

Going over the SwiftUI observation system

Before we discuss the current SwiftUI observation system, let's recap the SwiftUI observation system.

Before Xcode 15, nine property wrappers handled state and data updates in SwiftUI.

Let's try to group them by app levels:

- **Sub-View level**: `@Binding`, `@Environment`
- **View level**: `@State`, `@Binding`, `@StateObject`, `@Environment`
- **Business Logic level**: `@ObservableObject`, `@Published`
- **App/Data level**: `@AppStorage`, `@SceneStorage`, `@EnvironementObject`

The different levels give us an idea of the different roles of the different wrappers. Let's touch on some of these wrappers to understand how the system works.

A local `@State` property wrapper manages the state of primitive properties within the view. For example, whether a specific view is hidden, the number of available buttons, the current sorting method, and more are managed by this wrapper.

The reason why we use a `@State` property wrapper is because SwiftUI views are immutable. This means that SwiftUI rebuilds the view each time a change occurs, but the `@State` values don't change between one rendering session and another.

The problem begins when we base our view on data model information. An example of this would be a bookstore app that displays a list of books from a local data file. In this case, our view must work with another data model object using the `ObservableObject` protocol.

Let's go over it now.

Conforming to the ObservableObject protocol

We can use the `ObservableObject` protocol in conjunction with the `@ObservedObject` property wrapper for classes that need to be observed.

Here's an example of a `UserData` class which becomes an `@ObservedObject` property wrapper:

```
class UserData: ObservableObject {
    @Published var username = "Avi Tsadok"
}

struct ContentView: View {
    @ObservedObject var userData = UserData()
```

```
    var body: some View {
        Text("Welcome, \(userData.username)!")
            .padding()
    }
}
```

There are three parts to implementing a data class observation:

1. **Conforming to `ObservableObject`**: If we want a class to be observed in SwiftUI, it must conform to the `ObservableObject` protocol. This indicates to SwiftUI that any instance derived from this class can be observed in a view.
2. **Adding the `@Published` property wrapper**: When we mark a property with a `@Published` property wrapper, SwiftUI creates a publisher and uses it inside the SwiftUI views.
3. **Marking variables with the `@ObservedObject` property wrapper**: The `@ObservedObject` property wrapper establishes a connection between the view and the object, allowing the view to be notified of changes.

It's essential to remember that the `@ObservedObject` property wrapper is solely for observation purposes – this means that the view cannot modify the observed object properties directly.

If we want to change the observed object properties, we must use another property wrapper – `@StateObject`.

A `@StateObject` property wrapper is similar to `@State`, only that it works with observable objects and not primitive values.

However, that doesn't end here – if we want to create a two-way connection between the view and its subview, we need to add a `@Binding` property wrapper to the subview and a `@State` property wrapper to the parent view.

Explaining the problem with the current observation situation

The short recap of the current way observation works in SwiftUI emphasizes how complex and confusing it is to observe data in SwiftUI.

Take, for example, the `ObservableObject` protocol – in most cases, we want to mark all of our properties with the `@Published` property wrapper. If that's the case, why do we need to work hard? Don't we have a way to add a `@Published` property wrapper to all our properties?

The observation framework uses Swift macros here, a feature that can help us reduce boilerplate code. To read more about it, go to *Chapter 10* and read about Swift macros.

Adding the @Observable macro

The primary goal of the Observation framework is to simplify our work as much as possible, and it does that with the heavy use of macros.

Let's take a `Book` class, for example:

```
class Book: ObservableObject {
    @Published var title:String = ""
    @Published var author: String = ""
    @Published var publishedYear: Date = Date()
    @Published var numberOfPages: Int = 0
}
```

The `Book` class is a standard `ObservableObject` class that contains four properties with a `@Published` property wrapper.

Using the `Observation` framework, we can get rid of all the property wrappers and the `ObservableObject` protocol and just add a macro attached to the class declaration:

```
@Observable
class Book {
    var title:String = ""
    var author: String = ""
    var publishedYear: Date = Date()
    var numberOfPages: Int = 0
}
```

The `@Observable` macro, like most macros, handles the tedious work on our behalf. It makes the `Book` struct observable and adds a publisher to its properties.

Let's try to use the `Book` class in a view:

```
struct ContentView: View {

    var book:Book = Book()
    var body: some View {
        VStack {
            Text(book.title)
            Button("Change") {
                book.title = "Mastering iOS 17"
            }
        }
        .padding()
    }
}
```

In the preceding code, we have a button and a view with a `Text` view that displays the book title. Tapping on the button changes the book title.

The change of the book title updates the text; however, the updates happen even though the book is not marked with a `@ObserverdObject` or `@StateObject` property wrapper!

How can this be?

Let's dive a little bit deeper to find out!

Learning how the @Observable macro works

I know speaking about macros might get on your nerves, but do you remember that `@Observable` is a macro, and that we can expand it?

So, let's expand it and see what's going on there:

```
@Observable
class Book {
    @ObservationTracked
    var title:String = ""
        @ObservationIgnored private var _title: String = ""
    {
        @storageRestrictions(initializes: _title)
        init(initialValue) {
          _title = initialValue
        }

        get {
          access(keyPath: \.title)
          return _title
        }

        set {
          withMutation(keyPath: \.title) {
            _title = newValue
          }
        }
    }
    @ObservationTracked
    var author: String = ""
    @ObservationTracked
    var publishedYear: Date = Date()
    @ObservationTracked
    var numberOfPages: Int = 0
```

```
      @ObservationIgnored private let _$observationRegistrar
        = Observation.ObservationRegistrar()

      internal nonisolated func access<Member>(
        keyPath: KeyPath<Book , Member>
      ) {
          _$observationRegistrar.access(self, keyPath:
          keyPath)
      }

      internal nonisolated func withMutation<Member,
        MutationResult>(
        keyPath: KeyPath<Book , Member>,
        _ mutation: () throws -> MutationResult
      ) rethrows -> MutationResult {
        try _$observationRegistrar.withMutation(of: self,
          keyPath: keyPath, mutation)
      }
  }
  extension Book: Observation.Observable {
  }
```

That's a lot of work for one tiny macro!

It looks like there are also internal macros, such as `@ObservationTracked`, one of which I expanded.

So, what's going on here?

There are five things we can see here:

- **Our class conforms to the Observable protocol**: Don't be confused; we discussed a macro named `Observable`, not a protocol. The protocol itself is empty, but SwiftUI uses that to mark the class as observed. Using an extension, you can see the protocol conformance at the end of the macro code.

- **Having a reference to** `observationRegistrar`: The `observationRegistrar` variable is a singleton struct responsible for managing the registration of observed class properties. SwiftUI relies on this struct to detect when an observed property is accessed or modified.

- **Our variables became computed**: We now have a setter and a getter for each variable. The `Observation` framework needs these getters and setters to track every access or modification attempt.

- **Each variable has a private variable for storage**: Since our variables are now computed, we need to store the actual values somehow. Our `@Observable` macro added a private variable for each original variable just for that. The getter and the setter use the private variable to return and mutate the stored values.

- **Two private methods were added**: We now have the `access()` and `withMutation()` methods. The computed variables call these methods to notify the `observationRegistrar` instance about any data modification access. Afterward, the `observationRegistrar` instance tells SwiftUI about these changes.

The reason we have so much code underneath is that the *Observation* framework's goal is to simplify the process of observing data models. Conforming the class to the *Observable* protocol without the macro is not enough – marking the actual model with `@ObservedObject` would still be required in the SwiftUI view. The *Observation* framework tracks each property using its getter and setter methods, making it much cleaner to implement in our views.

Notice that there's a small macro inside that expanded code we haven't discussed – `@ObservationIgnored`.

Excluding properties from observation using @ObservationIgnored

We already understand that, unlike the previous pattern of adding the `@Published` property wrapper for each variable, in the `@Observable` macro, all the properties are observed by default.

Let's think of the consequences of that – how can it affect our work?

The fact that every property is now observed means that each time it appears in our SwiftUI view and we modify it, our view gets updated.

SwiftUI is indeed a highly optimized framework, but it is optimized because it only updates views when needed. If a particular data model property doesn't need to be dynamic and observed, we should exclude it from tracking. It's essential to balance observing many properties to keep our UI responsive and impact its performance.

Let's try to add a property that is not supposed to be observed:

```
@Observable
class Book {
    var title:String = ""
    var author: String = ""
    var publishedYear: Date = Date()
    var numberOfPages: Int = 0

    @ObservationIgnored
    var lastPageRead: Int = 0
}
```

In this code example, we added a property named `lastPageRead`. It's an important property, but it doesn't affect our UI state, and we don't display or even consider it when laying our views. Therefore, we will ignore it using the `@ObservationIgnored` macro.

Unlike the `@ObservationTracked` macro, which the `@Observable` macro uses to create the getters and the setters for the observed properties, `@ObservationIgnored` doesn't modify the property. SwiftUI uses that macro only to determine which property it doesn't register using the `observationRegister` object.

The default observation of all properties gives us another exciting and powerful feature out of the box – observing **computed variables**.

Observing computed variables

First, a reminder – a computed variable is a property that has a getter and an optional setter. This means that a computed variable doesn't have its storage, and its value is derived from other variables (which can also be computed variables).

Look at the following code:

```
class Book: ObservableObject {
    @Published var pages: Int = 0
    @Published var averageWordsPerPage: Int = 0
    @Published var totalWordsInBook: Int {
        return pages * averageWordsPerPage
    }
}
```

The `Book` class conforms to the good old `ObservableObject` protocol.

Notice that the `totalWordsInBook` property is a computed variable – it multiplies the `pages` and `averageWordsPerPage` variables to return the total number of words in the book.

We want to observe the computed variable to present its results in one of our SwiftUI views, so we have marked it with the `@Published` property wrapper.

Unfortunately, this is impossible. Try to compile results with the following error:

Property wrapper cannot be applied to a computed property

That's a big downside for conforming to the `ObservableObject` protocol, as it can be a helpful use case.

Working with the Observable macro works this out quite nicely:

```
@Observable
class MyBook {
```

```
    var pages: Int = 0
    var averageWordsPerPage: Int = 0
    var totalWordsInBook: Int {
        return pages * averageWordsPerPage
    }
}
```

In the preceding code, we just added the computed variable, and we can observe it in our view with no problems.

How does it work? How can we observe a computed variable if it doesn't have a back store for its value?

So, there's a reason why I always make sure to explain how things work underneath. If we go back to the *Learning how the @Observable macro works* section, we expanded the @Observable macro and saw interesting details of how the observation and tracking work. Every observed property becomes a computed value and is tracked using a getter and a setter.

So, when we add a computed variable whose value is derived from another observed property, it means that whenever we access this computed variable, we also access the other properties. This access triggers the observation framework.

Figure 3.1 shows how observing computed variables works in a visual way:

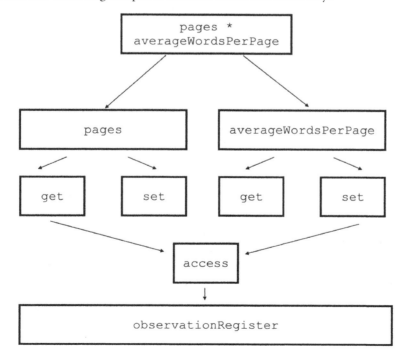

Figure 3.1: How SwiftUI observes computed variables

Figure 3.1 nicely shows how the computed variables are derived from other properties and how accessing them would eventually go down to the `observationRegister` object.

Let's try to see that in action:

```
@Observable
class Book {
    var title:String = ""
    var pages: Int = 0
    var averageWordsPerPage: Int = 0
    var totalWordsInBook: Int {
        return pages * averageWordsPerPage
    }
}

struct ContentView: View {
    var book:Book = Book()
    var body: some View {
        VStack {
            Text(book.title)
            Button("Change") {
                book.averageWordsPerPage = 300
                book.pages = 200
            }
            Text("number of pages in the book:
              \(book.totalWordsInBook)")
        .padding()
        }
    }
}
```

In the preceding code, we update the `averageWordsPerPage` and `pages` properties when tapping the **Change** button.

The update triggers the observation framework and updates the view because we access `totalWordsInBook` in the following line, even though it's a computed variable.

However, adding the `@ObservationIgnored` property to both of these properties (`averageWordsPerPage` and `pages`) won't trigger the `totalWordsInBook` computed property because the `@Observation` framework can't tell that something has changed. The nice thing is that we've learned how it works by expanding our `@Observable` macro.

By now, we know very well how the `@Observable` macro works and how variables and computed variables are observed.

Now, let's move one step further and see how to use these observed variables as environment variables.

Working with environment variables

A view that works directly with an observed object is a common use case. For example, a view can work with a `ViewModel` class or have a SwiftData query that fetches data models from the persistent store.

However, there are cases where we have an observed object shared across different views.

Some examples of such a use case are as follows:

- **App settings**: The user profile is part of app settings and can be stored in an environment variable
- **Themes and styling**: Primary color tint font style, spaces, and more
- **User authentication state**: The login state is a good example of an environment variable

Sharing the same object down a view hierarchy can be cumbersome, but SwiftUI offers a helpful feature known as **environment variables**. While environment variables aren't a recent addition to iOS (they have been available before iOS 17), the Observation framework provides comprehensive support.

There are two ways of adding an environment variable to our project – by type or by keys. Let's start with the more straightforward way: by type.

Adding an environment variable by type

Let's try to add theming support for our books project. We'll start by creating our `Themes` class:

```
@Observable
class Themes {
    var primaryColor: Color = .red
}
```

Our `Themes` class has only one property for now: the primary color. Notice that we added the `@Observable` macro to update our UI when the theme changes.

Next, we will add our observed object to our `BookApp` struct:

```
@main
struct BookApp: App {
    var themes: Themes = Themes()
    var body: some Scene {
        WindowGroup {
            ContentView()
                .environment(themes)
        }
    }
}
```

In the `BookApp` struct, we're making two changes:

- **Creating a new Themes object**: That object stores the current value theme value and state. Notice that we don't need to mark it as `@State` or `@ObservedObject`.
- **Adding the environment modifier**: The environment modifier allows the child views to make use of the `themes` object easily.

Now, let's turn to our view and see how we can use it:

```
struct ContentView: View {
    @Environment(Themes.self) var themes
    var book: Book = {
        let book = Book()
        book.title = "Mastering iOS 17"
        return book
    }()
    
    var body: some View {
        VStack {
            Text(book.title).foregroundStyle(themes.primaryColor)
        }
    }
}
```

Adding the themes instance to our `ContentView` struct is straightforward. We're using the `@Environment` property wrapper to inject the themes object we created earlier.

We use the theme's primary color in the body part to color our book title.

Now, we must note that we can use the environment variable in *every* view in the hierarchy, even if we haven't initialized it with the environment modifier.

Here's an example of that:

```
struct ContentView: View {
    var body: some View {
        VStack {
            MyTitle(text: "Mastering iOS 17")
        }
    }
}

struct MyTitle: View {
    @Environment(Themes.self) var themes
    let text: String
```

```
    var body: some View {
        Text(text).foregroundStyle(themes.primaryColor)
    }
}
```

In the preceding code, we created another SwiftUI component called `MyTitle`, which has the environment variable themes.

The `MyTitle` view is part of the `ContentView` hierarchy. Therefore, it has direct access to the `themes` variable.

Passing environment variables by type is simple! However, when working on a big scale, it has some drawbacks. I believe that the main disadvantage is that we are coupling our code to a specific type. In the `themes` example, we work with an explicit variety (`Themes`).

SwiftUI provides us with a better way to manage environment variables, and that's working with environment keys.

Adding environment variable by key

Managing environment variables is much better when our project becomes more significant.

Using environment keys improves the separation between our view and the actual variable.

To better manage environment values, SwiftUI has two primary components:

- `EnvironmentValues` **struct**: This is a container of different environment values structured in a key-value form. It can be accessed from any view in the app. We can extend the struct and add new variables.
- `EnvironmentKey` **protocol**: It allows us to add a key for a new variable and use that key to add a new environment value.

Let's see how it works in practice:

```
struct ThemesKey: EnvironmentKey {
    static let defaultValue = Themes()
}

extension EnvironmentValues {
    var themes: Themes {
        get { self[ThemesKey.self] }
        set { self[ThemesKey.self] = newValue }
    }
}
```

The first thing that we did was add a new `EnvironmentKey` type named `ThemesKey`. Part of the `EnvironmentKey` protocol is setting the variable default value, which is, in this case, a `Themes` instance.

Once we have a new environment key, we must add it to our `EnvironmentValues` container. We do that by extending the container and adding a new computed variable named `themes`.

The getter and the setter are straightforward – the `get` function returns the value according to the relevant key (`ThemesKey`), and the `set` function stores a new variable on that key.

After we have extended the container, we can easily access that key from any view that we have:

```
struct ContentView: View {
    @Environment(\.themes) var themes
    // rest of the view
}
```

Remember the environment modifier from earlier? We can remove it now:

```
ContentView()
    .environment(themes)
```

When we extended the `EnvironmentValues` struct, we extended the global variables container of our app. That's the reason why we have access from any view.

Other than accessing the values from any view, working with environment variable keys has several additional advantages:

- **Quickly replacing the variable type in the future**: Unlike adding an environment value by type, we are not tied to a specific type when adding the variable by key. We can easily replace the type itself in one place and not have to replace it in all views as long as we keep the same interface.
- **Great for testing**: Another advantage of not being coupled to a specific type is the ability to create mocks and add unit tests.
- **Adding more flexibility**: Since the instance creation is not in the view hierarchy, it is easier to have more control. Remember the `get` and `set` functions in the `EnvironmentValues` struct? Now, we can customize them the way we want to.

We can understand why environment keys are essential for big projects by looking at the list of advantages.

No matter how we work with environment variables, they are crucial for a clean and simple SwiftUI code, especially when we combine them with `@Observable` objects.

By now, we already know how to create an observed object and inject it into child views using environment variables.

Our next topic revolves around the compatibility problem that the *Observation* framework created for us, specifically regarding binding.

Binding objects using @Bindable

Let's start with a short recap of what binding is.

In some cases, a view and its child must share a state and create a two-way connection for reading and modifying a value. To do that, we use something called **binding**.

One classic example is `TextField` – a `TextField` view is a SwiftUI component with a `text` variable. Both `TextField` and its parent view share the same value of text. Therefore, it's a binding variable:

```
struct ContentView: View {
    @State var email: String = ""
    var body: some View {
        VStack {
            TextField("Email", text: $email)
        }
    }
}
```

We see that the `email` variable is marked as a state, but the `TextField` view is the one that updates it. The binding occurs using the `$` character.

We can create a binding variable ourselves using the `@Binding` property wrapper:

```
struct MyCounter: View {
    @Binding var value: Int
    var body: some View {
        VStack {
            Button("Increase") {
                value += 1
            }
        }
    }
}

struct ContentView: View {
    @State var count: Int = 0
    var body: some View {
        VStack {
            MyCounter(value: $count)
            Text("Value = \(count)")
        }
```

```
        }
    }
```

The `count` variable in the parent view (`ContentView`) and the `value` variable in the child view (`ContentView`) share the same state, and now we have a two-way connection between them.

We can connect a binding variable to a `@State` property wrapper (such as in the example we just saw) or a `@ObservedObject` variable.

Can you guess what the problem is with trying to create a binding connection using the Observation framework?

So, apparently, classes that are marked with the `@Observed` macro are not eligible for `@State` or `@ObservedObject`, so we can't use `@Binding` with them.

Fortunately, with the *Observation* framework, we have a new property wrapper called **@Bindable**.

Let's see a short example of how to use `@Bindable` with a counter object:

```
struct ContentView: View {
    var counter = Counter()

    var body: some View {
        VStack {
            CounterView(counter: counter)
            Text("Value = \(counter.value)")
        }
    }
}
struct CounterView: View {
    @Bindable var counter: Counter

    var body: some View {
        VStack {
            Button("Increase") {
                counter.increment()
            }
        }
    }
}
```

The code example has two views as before – a `ContentView` view and a child view named `CounterView`. The `ContentView` view has a variable called `counter` of the `Counter` type. The `Counter` class is marked with `@Observed`, so we don't need to mark the property as `@State` or `@ObservedObject`.

In the `CounterView` structure, we also have a counter from the same type, but it is marked with `@Bindable`. This means we need to bind it to an object with a similar type.

The `CounterView.counter` and `ContentView.counter` variables are linked – whenever we change the value in the child view, it automatically reflects in the parent view. Notice that with `@Bindable`, we don't need to add any `$` signs to the variable expression. Everything just works.

Binding is a critical usage of SwiftUI – it stands at the heart of many input views such as text fields, toggles, sheets, and more.

Working with the `@Bindable` macro can be confusing – we now have both `@Binding` and `@Bindable` at the same time! `@Binding` is used for states and observable objects and `@Bindable` is used for... observed objects?

So yes, it feels like we are in a transition era. The good news is that we can solve the issue easily by migrating our project to *Observable*.

Migrating to Observable

Before migrating to *Observable*, we must ensure that our app deployment target is at least 17. Remember that this feature (and most of the new features described in this book) are from iOS 17, and some are irrelevant if our app deployment target is not 17.

Let's try to recap the different Observable attributes:

- `@State`: This is used to manage the state within a specific view. A change to a `@State` property triggers a view update. For example, data related to a list or view visibility can be marked as `@State`.
- `@Observable`: This can be applied to a class to make the class observable. Each class property is automatically marked with `@Published` unless we mark them as `@ObservataionIgnored`. `@Observable` can be added to view models or business logic classes.
- `@Bindable`: This creates a two-way connection between a property and another value. Text field input, toggles, or a counter are examples of views for implementing a `@Bindable` connection.
- `@Environment`: Mark an object to be shared down the view hierarchy with this attribute. For example, configuration or a theme can be shared with all views in the hierarchy using the `@Environemnt` attribute.

This list aims to summarize the different attributes in the Observable framework and their use cases.

Once we decide to move to the *Observable* framework, there are a few things we need to do:

- Remove the protocol conformance to `ObservableObject` and add the `@Observable` macro for all the relevant classes

- Remove the `@Published` property wrapper and add `@ObservationIgnored` for the properties we don't want to observe
- Remove the `@ObservedObject` property wrapper
- Rename `@Binding` to `@Bindable` for the properties that are based on classes

Once we finish migrating to the `Observable` framework, things will be clearer and more straightforward, with fewer property wrappers and less protocol conformation. The binding can also be simple – now it's `@Binding` for primitive values and `@Bindable` for classes. That's not perfect, but not too bad either. It's time to enjoy *Observable*!

Summary

This was another chapter that made use of Swift macros and other advanced Swift techniques. A small note: to understand topics such as *Observable*, I recommend having good knowledge of Swift. Otherwise, it becomes just another boring tutorial. Knowing how things work on the inside is fascinating and can only make us better.

In this chapter, we did a recap of the SwiftUI observation system, and we discussed its problem. We added the `@Observable` macro and explored how it works. We talked about computed variables, environment variables, and bindable. Ultimately, we discussed migrating from the "old" observation system to the new *Observable* framework.

Remember – observation is a core feature of SwiftUI and is crucial to delivering a superior experience to our users.

In the next chapter, we will learn about another critical feature, especially in mobile – navigation and search.

4
Advanced Navigation with SwiftUI

In *Chapter 2*, we discussed working with the **Observation** framework. The Observation framework helps to manage communication between different parts of our app and is one of the fundamental building blocks of SwiftUI declarative programming. However, it is also one of the tools we will use to implement a good navigation system.

Why do we have a whole chapter about navigation? Isn't it just showing a different view when the user selects an item in a list?

Navigation is a massive topic in mobile development. A standard app may have dozens of screens, and a more extensive one may have hundreds. Understanding how to manage the different routes in our app, which has so many screens, is crucial to our app's success.

In this chapter, we will be doing the following:

- Understating why SwiftUI navigation is a challenge
- Exploring SwiftUI's `NavigationStack`
- Working with different data models to trigger navigation
- Working with the Coordinator pattern to manage our concerns better
- Implementing SwiftUI's `NavigationSplitView` to create a column-based navigation

We've got a lot to cover! But before we begin, let's try to understand why SwiftUI navigation can be a challenge.

Technical requirements

For this chapter, you'll need to download Xcode version 16.0 or above from Apple's App Store.

You'll also need to be running the latest version of macOS (Ventura or above). Simply search for Xcode in the App Store and select and download the latest version. Launch Xcode and follow any additional installation instructions that your system may prompt you with. Once Xcode has fully launched, you're ready to go.

Download the sample code from the following GitHub link: `https://github.com/PacktPublishing/Mastering-iOS-18-Development/tree/main/Chapter%204`.

Understating why SwiftUI navigation is a challenge

To answer that question, we need to understand how navigation works intuitively. The user taps on a button, link, or some other event that may occur. Then, the app responds to that event and transitions the view to another screen.

In a sense, we understand this sounds like an event-driven paradigm. When we discuss the differences between SwiftUI and UIKit, we actually discuss the differences between declarative and imperative programming.

Imperative UI, such as UIKit, is also event-driven, while declarative UI, such as SwiftUI, represents the current state. As a result, we can understand why navigation can be seen as simpler in UIKit and may feel more natural there.

Many developers struggle with SwiftUI navigation. They wrap a SwiftUI view inside `UIHostingController` and use the UIKit navigation system. That's a fair solution for achieving some advanced navigation techniques that are hard to do in SwiftUI. However, we need to remember that SwiftUI has evolved over the years and offers great navigation tools.

Let's start with the basic navigation tool – `NavigationStack`.

Exploring NavigationStack

When SwiftUI was introduced, the basic navigation mechanism was based on a view called `NavigationView`. However, `NavigationView` was too simple for most apps, and `NavigationStack` replaced it. In fact, Apple deprecated `NavigationView`, starting with iOS 18.

Compared to `NavigationView`, `NavigationStack` adds a little bit of complexity to the pile, which provides us with new capabilities.

Let's see a simple example of a `NavigationStack` usage:

```
struct ContentView: View {
    var body: some View {
        NavigationStack {
            NavigationLink("Tap here to go to the next
            screen") {
                Text("Next Screen!")
```

```
            }
         }
      }
}
```

This code example looks pretty simple!

However, `NavigationStack` is much more powerful than it seems.

How? The concept of `NavigationStack` is constructed from four components:

1. **Separate NavigationLink from its destination**: This is a major change from `NavigationView`. In `NavigationStack`, `NavigationLink` describes what happened, and the `navigationDestination` view modifier describes where we go.
2. **Linking between data and destinations**: In a way, this is a development of the preceding point. The destination is linked to a data type. This means that we can have several navigation links that point to the same destination just because they share the same data type.
3. **Allowing us to read and update the path**: Here, we have another development of our idea. Because the data and the screen are now linked, we can represent the path as an array of data instances. Modifying the path array also changes our views stack.
4. **Presenting views without navigation links**: Prior to iOS 16, `NavigationLink` also had this capability, but the introduction of `NavigationStack` made it obsolete.

Let's cover each of these four components in detail now, and we'll start with destinations.

Separating the navigation destination using the navigationDestination view modifier

If you've read my previous books (*Pro iOS Testing* and *Mastering Swift Package Manager* by *Apress*, and *The Ultimate iOS Interview Playbook* by *Packt Publishing*) there's an important principle I keep nagging about: **separation of concerns** (**SoC**). In SoC, we break our code into distinct modules or components, each with a specific and well-defined responsibility. This makes our code more modular, flexible, and easy to maintain.

When we look back at `NavigationLink`, we can see that it has more than one responsibility – it is the actual control that the user taps on and also contains the next screen view.

In `NavigationStack`, there's a new view modifier called `navigationDestination`, which allows us to define a destination separately according to a state change.

Let's see an example of `navigationDestination`, based on a binding variable:

```
struct ContentView: View {
    @State var isNextScreenDisplayed: Bool = false
```

```swift
var body: some View {
    NavigationStack {
        Button("Go to next screen") {
            isNextScreenDisplayed = true
        }
        .navigationDestination(isPresented:
          $isNextScreenDisplayed) {
            Text("Next Screen!")
        }
    }
}
```

In our code example, we can see a `NavigationStack` view containing a button. Notice that there's no `NavigationLink` view at all, and that's because we don't need it. We trigger the navigation by changing the `@State` property named `isNextScreenDisplayed` rather than using a `NavigationLink` view.

The button also has a view modifier called `navigationDestination`. The `navigationDestination` view modifier has a binding Boolean variable that is linked to the `isNextScreenDisplayed` state variable. It also has a view builder that contains our next screen (similar to `NavigationLink`).

Tapping on the button toggles `isNextScreenDisplayed` and navigates our next screen.

This capability of triggering navigation using `NavigationLink` was available in earlier versions of SwiftUI, but it is deprecated now. But don't worry – decoupling the destination from the actual control makes our code much more flexible and provides us with more opportunities.

For example, imagine we're doing an asynchrony operation such as a network request or image processing, and we want to move to the next screen – that can be done easily by toggling a Boolean variable.

Another important aspect of having a separate destination is that we can trigger the same navigation from different places. We can toggle the Boolean from an asynchronous operation and a button as well. Responding to a state follows a declarative approach rather than the `NavigationView` approach, which was responding to a button tap.

Toggling a Boolean variable is great when navigating to a new screen unrelated to any data. For example, moving to the settings from our main screen is a classic example of using a Boolean binding.

But I promised that `NavigationStack` has more than that, didn't I?

So, let's see how we can bind our navigation destinations to data models.

Using data models to trigger navigation

Developers who are used to UIKit navigation may find the idea of using data models weird. After all, toggling a Boolean for navigation is quite similar to imperative programming, but how does a data model have anything to do with navigation?

We understand that many screens are related to a specific data model. For example, tapping on a movie leads us to a single movie screen if we have a list of movies. Another example is a trips app, where tapping on a specific trip leads us to a screen dedicated to that trip.

If we think even deeper than that, we can represent many screens in our app using a data model. We can distinguish between screens using a data model containing an enum.

Before we set sail with our thoughts, exploring potential possibilities and implementations, let's see what basic data-based navigation looks like:

```
struct ContentView: View {
    private let countries = ["England", "France", "Spain",
    "Italy"]

    var body: some View {
        NavigationStack {
            List(countries, id: \.self) { country in
                NavigationLink(country, value: country)
            }
            .navigationDestination(for: String.self) { item
                in
                Text(item)
            }
        }
    }
}
```

As always, I have highlighted the interesting parts in the preceding code. We have a SwiftUI view that displays a list of countries (based on a constant variable).

Each row has a `NavigationLink` view that displays the country name, but it doesn't have its own destination this time. Instead, it uses the country as the link's value parameter.

We can understand what sending the country as a value means only when we look down at the navigation destination. In the code example in the *Separating the navigation destination using the navigationDestination view modifier* section, the navigation destination was linked to a Boolean state variable. In this case, the navigation destination performs only when there's a link with a specific data type – in this case, a string type (just like a country value).

In other words, tapping on a country sends its value to the navigation stack using `NavigationLink`. The navigation destination catches that and defines what will be our next screen.

We can use the data models to navigate to different places by defining multiple navigation destinations, each responding to a different data model type.

Here's another example of using a navigation destination to add a navigation to a profile screen:

```swift
struct Profile: Hashable {
    let firstName: String
    let lastName: String
    let email: String
}

struct ContentView: View {
    
    let profile = Profile(firstName: "Avi", lastName: "Tsadok", email: "myemail@domain.com")
    
    let countries = ["England", "France", "Spain", "Italy"]
    
    var body: some View {
        NavigationStack {
            List(countries, id: \.self) { country in
                NavigationLink(country, value: country)
            }.toolbar(content: {
                NavigationLink("Go to profile", value: profile)
            })
            
            .navigationDestination(for: String.self) { item in
                Text(item)
            }
            .navigationDestination(for: Profile.self) {
                profile in
                VStack {
                    Text(profile.firstName)
                    Text(profile.lastName)
                    Text(profile.email)
                }
            }
        }
    }
}
```

In the preceding code, we see another navigation destination for a data model from the type of `Profile`. To navigate the profile screen, we added another `NavigationLink` view in the screen toolbar and sent the profile data model.

Our navigation system is dynamic because we can work with different data models. But that doesn't stop here. `NavigationStack` can also reveal and even modify the current view's stack. We do that using the path binding variable.

Responding to the path variable

Separating the destination from its navigation link is great, but `NavigationStack`'s ability to observe and update its stack of views is very powerful.

As mentioned, a `NavigationStack` view has a binding variable called `path`, and the `path` variable can contain the list of views by their data models.

It is easy to demonstrate that using a linked list:

```
struct ContentView: View {
    let list: LinkedList<Int> = {
        let list = LinkedList<Int>()
        list.head = ListNode(1)
        list.head?.next = ListNode(2)
        list.head?.next?.next = ListNode(3)
        return list
    }()

    @State var path: [ListNode<Int>] = []

    var body: some View {
        NavigationStack(path: $path) {
            VStack {
                NavigationLink("Start", value: list.head)
            }
            .navigationDestination(for: ListNode<Int>.self)
              { node in
                NavigationLink("\(node.value)", value:
                node.next)
              }
        }
    }
}
```

I chose to demonstrate working with `path` using a linked list since it's a great data structure that is similar to a navigation stack (linked items from the same type).

If we observe the `path` variable during navigation, we can see it contains a collection of the list nodes currently active as views.

What's great about the fact that the `path` variable is bound to the `NavigationStack` is that we can manipulate and modify it:

```
path.append(ListNode(4))
```

Appending a new list node to `path` triggers the navigation and directs the user to a new screen.

We can also create a whole stack using the `path` variable:

```
path = [ListNode(1), ListNode(2)]
```

Setting a new array of nodes creates a new stack of corresponding views. This is a great way to implement a deep link or direct the user to a specific location within the app.

You are probably scratching your head right now and thinking, how can we implement it inside an app? What are the use cases where we navigate a few levels down the hierarchy with the same data model type?

So, a data model type doesn't have to be `Task`, `Album`, or `Contact`. A data model can also describe a screen or a feature. In this way, data collection can describe a navigation path inside an app.

Here's an example of a data type that can describe a screen, followed by a navigation path:

```
enum Screen: Hashable {
    case signin
    case onboarding
    case mainScreen
    case settings
}

@State var path: [Screen] = []
```

The Enum `Screen` describes the type of screen we want to navigate to, and it's an easy way to build a stack:

```
path = [.mainScreen, .settings]
```

That short line of code builds a stack of views when the first view is the main screen followed by a settings screen.

Using an Enum to display different kinds of screens is great. However, working with different types of data is less convenient with Enum. To solve that issue, we have a more complex solution than a collection of instances, and it's called `NavigationPath`.

Working with different types of data using NavigationPath

`NavigationPath` was introduced along with `NavigationStack`, and it allows us to have more control of our navigation flows. In fact, `NavigationPath` makes navigation with SwiftUI a mix of declarative and imperative programming and is much more similar to the UIKit navigation pattern.

Let's say we have a music app with a list of songs and albums on its main screen. Tapping on a song leads to a `song` view while tapping on an album navigates to an `album` view. In the previous section, we managed that using an Enum, trying to map the Enum value to a screen view. With `NavigationPath`, we can append whatever value we want to the `path` variable as long as its type conforms to `Hashable`.

Let's have a look at the following code:

```swift
struct ContentView: View {
    @State private var navigationPath = NavigationPath()
    @State private var albums: [Album] = [Album(title:
        "Album 1"), Album(title: "Album 2")]
    @State private var songs: [Song] = [Song(title: "Song
        1"), Song(title: "Song 2")]
    
    var body: some View {
        NavigationStack(path: $navigationPath) {
            VStack {
                List {
                    Section(header: Text("Songs")) {
                        ForEach(songs) { song in
                            Button(action: {
                                navigationPath.append(song)
                            }) {
                                Text(song.title)
                            }
                        }
                    }
                    
                    Section(header: Text("Albums")) {
                        ForEach(albums) { album in
                            Button(action: {
                                navigationPath.append(album)
                            }) {
                                Text(album.title)
                            }
                        }
                    }
                }
```

```
        }
        .navigationDestination(for: Song.self) {
    song in
        SongDetailView(song: song,
            navigationPath: $navigationPath)
    }
    .navigationDestination(for: Album.self) {
    album in
        AlbumDetailView(album: album)
    }
}
```

Note that the preceding code example is partial and does not include the child views.

Our music app's main screen contains four important parts that handle our navigation system:

We start with declaring a state variable that holds our `path` variable called `NavigationPath`:

```
@State private var navigationPath = NavigationPath()
```

As mentioned earlier, unlike the previous `path` variable we used, in the `NavigationPath` case, we don't need to define its type. It can hold any type we want as long as it conforms to `Hashable`.

Next, we will initiate `NavigationStack` with our new `NavigationPath` similar to what we did in the previous example:

```
NavigationStack(path: $navigationPath) {
```

Notice that we use a similar signature but with a different type – `Binding<NavigationPath>` instead of `Binding<Data>`.

Now that we have `NavigationPath`, we can navigate to a `song` view or to an `album` view by appending the corresponding object to the navigation path:

```
navigationPath.append(song)
```

Or, you can do it like so:

```
navigationPath.append(album)
```

The appending operation triggers the `navigationDestination` view modifier, passing the song or the album that was selected:

```
.navigationDestination(for: Song.self) { song in
    SongDetailView(song: song, navigationPath:
        $navigationPath)
}
```

```
.navigationDestination(for: Album.self) { album in
    AlbumDetailView(album: album)
}
```

In this example, we have a different `navigationDestination` view modifier for each type we pass.

The fact that we can append any entity we want makes `NavigationPath` an ideal component for a flexible navigation system.

We can also use `NavigationPath` to perform a `Back` operation by removing the last component:

```
Button("Back") {
    navigationPath.removeLast()
}
```

In this example, we added a back button that removes the navigation path's last components when tapped.

Because we are still in a declarative world, any change we perform to the navigation stack by appending or removing components reflects the change in our UI.

Working with the Coordinator pattern

The `NavigationPath` and `NavigationStack` combination is robust and provides flexibility in managing navigation. However, as our app scales, controlling how the user moves from screen to screen becomes more complex.

For example, let's say we have an onboarding flow and want a different set of screens for different user profiles. Or, we want to reuse the same screen within different flows. In each flow, the screen should continue to a different screen.

In each case, it becomes difficult to understand our next view when we are within the screen context. In fact, this problem of managing our navigation is not related only to SwiftUI, and most developers know that from UIKit.

To try and improve our navigation mechanism, we can use what's called a **Coordinator pattern** – a pattern that delegates the navigation logic to a dedicated component.

Let's try to understand what it means.

Understanding the Coordinator's principles

Before we write our first Coordinator together, let's review some fundamental principles:

- The Coordinator is a component that holds the current navigation path and general context. It knows what screen is displayed and the general current flow. The Coordinator also adds a new view to the stack, pops, and shows modal or sheet views.

- A view doesn't know the following view the user should navigate to. What it does know is only the action the user performed. In a way, the view is isolated from the navigation logic and is unaware of the general context.
- A coordinator represents a flow. We can have several flows in our app with several coordinators.

As a result of these principles, we can understand that the Coordinator pattern is an improved way of separating our app concerns.

Look at *Figure 4.1*:

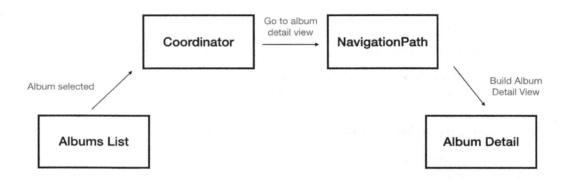

Figure 4.1: The Coordinator pattern

Figure 4.1 shows a basic Coordinator pattern. We have an *Albums List* view, and when the user selects an album, the action is sent to the *Coordinator*. Then, the Coordinator decides to navigate to the *Album Detail* view by sending the action to `NavigationPath`.

In this pattern, the Albums List is unaware of what should happen next. For example, the Coordinator can decide that, in some cases, we should show the user an upsell screen. Or, if it's part of onboarding, the Coordinator can determine that the Albums List is just a demonstration and that we should proceed to the next step in the onboarding flow.

But how do we structure a Coordinator pattern? How does it work, especially in the SwiftUI world?

There are many ways to build a Coordinator in SwiftUI. The Coordinator pattern I describe here is just an example that demonstrates the basic principles, and we can take that example and adjust it to our project's needs.

We will start with the most fundamental component – the Coordinator itself.

Building the Coordinator object

The Coordinator is the central object that defines the different user actions and navigation options. It also holds the navigation path so it can perform the navigation operations.

We will start by defining a basic Coordinator class:

```
class Coordinator: ObservableObject {
    @Published var path = NavigationPath()
}
```

We created a Coordinator class that holds a `NavigationPath` object. The `NavigationPath` object is essential – it allows the Coordinator to add more items to the stack, perform pop operations, and understand the current stack. Notice that the Coordinator conforms to the `ObservableObject` protocol and that the path is a published object – that's because we want the path to be part of `NavigationStack` when we use it.

Next, we define the different user and page actions:

```
enum PageAction: Hashable {
    case gotoAlbumView(album: Album)
    case gotoSettingsView
}

enum UserAction {
    case albumTappedInAlbumsList(album: Album)
    case settingButtonTapped
}
```

In this example, we created two Enums:

- `PageAction`: This Enum describes a navigation action our Coordinator needs to perform, such as navigating to an `album` view or a settings view.
- `UserAction`: This Enum describes an action the user performed, such as tapping on an album in the Albums List or tapping on the settings button.

Notice that some Enums contain associated values, such as the related `album` object.

Now that we have our Enums, we will create two important functions:

```
        func performedAction(action: UserAction) {
            switch action {
            case .albumTappedInAlbumsList(let album):
                path.append(PageAction.gotoAlbumView(album:
                album))
```

```
        case .settingButtonTapped:
            path.append(PageAction.gotoSettingsView)
        }
    }

    @ViewBuilder
    func buildView(forPageAction pageAction: PageAction) ->
      some View {
        switch pageAction {
        case .gotoAlbumView(let album):
            AlbumDetailView(album: album)
        case .gotoSettingsView:
            SettingsView()
        }
    }
}
```

The first is the `performAction()` function. This function receives `UserAction` as a parameter and appends the corresponding page action to `NavigationPath`. This function is the Coordinator's "brain" – where we decide where to navigate when the user performs a particular action.

In this example, when the user taps the album in the Albums List, we navigate to the `album` view, passing the `album` object. When the user taps the settings button, we navigate to the settings screen. This logic may sound evident and like over-engineering. Still, in a complex world, we have permissions, A/B tests, and other changes, and a centralized place that handles all of these can be extremely valuable.

The second function maps a page action to a SwiftUI view. We will use that now when we build `CoordinatorView`.

Adding CoordinatorView

The Coordinator class is robust and contains all of our navigation logic. However, we can't use the Coordinator to perform the actual navigation. To do that, we must wrap our views with `CoordinatorView`, which knows how to work with our Coordinator.

So, let's see what `CoordinatorView` looks like:

```
struct CoordinatorView: View {

    @ObservedObject private var coordinator = Coordinator()

    var body: some View {
      NavigationStack(path: $coordinator.path) {
        AlbumListView()
          .navigationDestination(for:
            PageAction.self, destination: { pageAction in
```

```
                    coordinator.buildView(forPageAction:
                       pageAction)
                 })
            }
            .environmentObject(coordinator)
       }
  }
```

`CoordinatorView` is a simple SwiftUI view that has three components:

- `coordinator`: In the `CoordinatorView`, we added an instance of our `Coordinator` class that we had just built. We made that coordinator an observable object so we can use its path to add and remove views from the stack.
- `NavigationStack`: This is the same `NavigationStack` we met in this chapter. As mentioned, we use the coordinator path as `NavigationStack`, but more importantly, two additional things – we initialize the stack with the root view (`AlbumListView`), and we use the Coordinator `buildView` function that maps the page action to view to add the corresponding view to the stack.
- `EnvironmentObject`: We add an `environmentObject` view modifier to declare an environment object in the coordinator. We do that to provide all the views under `NavigationStack` with access to the Coordinator so they can call the different user actions.

These three components are responsible for connecting our views to the Coordinator logic we have built.

Now, let's see how `AlbumListView` works with the Coordinator.

Calling the coordinator straight from the view

Remember one of our coordinator principles – the view's concern is only to say what *happened*, not what *will happen* next. What will happen is the Coordinator's concern.

Let's have a look at how `AlbumListView` deals with it:

```
struct AlbumListView: View {
    @EnvironmentObject private var coordinator: Coordinator

    var body: some View {
        List(albums) { album in
            VStack(alignment: .leading) {
                Text(album.title)
                    .font(.headline)
                Text(album.artist)
                    .font(.subheadline)
            }
```

```
            .onTapGesture {
                coordinator.performedAction(action:
                    .albumTappedInAlbumsList(album: album))
            }
        }
        .navigationTitle("Albums")
        .toolbar {
            ToolbarItem(placement: .navigationBarTrailing)
            {
                Button(action: {
                    coordinator.performedAction(action:
                        .settingButtonTapped)
                }) {
                    Image(systemName: "gear")
                }
            }
        }
    }
}
```

The `AlbumListView` struct contains a list of the user albums and a navigation bar with a settings button.

Tapping on one of the albums calls the Coordinator's `performedAction()` function, which returns the corresponding Enum and the selected album.

In addition, tapping on the settings button calls the same `performedAction()` function with a different Enum value.

Returning to the beginning of this section under the *Building the Coordinator object* part, we can now understand how everything is connected.

We can also understand why the coordinator instance is an environment object – so we can call it straight from the view.

Until now, we discussed `NavigationStack` and the Coordinator pattern. We might think that navigation is only about changing the current view. However, navigation on big screens, such as an iPad screen, often involves working with different columns. So, let's meet `NavigationSplitView` to see how we nail that down (I told you that navigation is a complex topic, didn't I?).

Navigating with columns with NavigationSplitView

One of the things that we know when building apps for padOS or macOS is that we need to take advantage of the big screen. But what does it mean? Sometimes, it might mean working with a grid instead of a list. However, in the context of navigation, it means that we can work with several columns when each of the columns shows a different view instead of replacing the whole screen each time the user navigates.

In other words – we need to split the screen.

To do that, we can work with a view called `NavigationSplitView`, which presents views in two or three columns.

When a user selects an item of one view, it updates the view in the other columns.

Creating NavigationSplitView

To demonstrate how to use `NavigationSplitView`, we will use our music app example and adjust it to padOS.

Let's start with some important terms – we have three different column types:

- `Sidebar`: The first column from the left. That's the main column where we start our navigation.
- `Content`: When there are three columns, the `Content` column shows data related to the selected item in the `Sidebar` column.
- `Detail`: The `Detail` column presents the selected item in the `Content` column or the `Sidebar` column. In general, it is the item that is last in the split view hierarchy.

These three terms may initially sound slightly confusing, so let's jump straight to the code to understand how they all fit together. Here's an example of `NavigationSplitView` that shows a list of albums, and when tapping on an album, the app shows a list of its songs:

```
var body: some View {
    NavigationSplitView {
        List(albums, selection: $selectedAlbum) { album
            in
            NavigationLink(album.title, value: album)
        }
    } detail: {
        if let selectedAlbum = selectedAlbum {
            List(selectedAlbum.songs, selection:
              $selectedSong) { song in
                Text(song.title)
            }
            .navigationTitle(selectedAlbum.title)
        } else {
            Text("Select an album")
        }
    }
}
```

Our code shows `NavigationSplitView` with two parts – the sidebar (the first block) and the detail. The sidebar shows a list of albums. Tapping on an album updates the `selectedAlbum` state variable. The `detail` block presents a list of songs about the selected album.

Let's see how it looks on an iPad within landscape orientation (*Figure 4.2*):

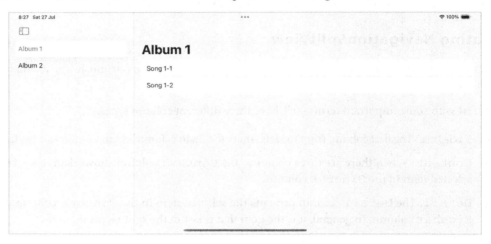

Figure 4.2: Two columns in SplitView on iPad – landscape

Here is how it appears in portrait orientation (*Figure 4.3*):

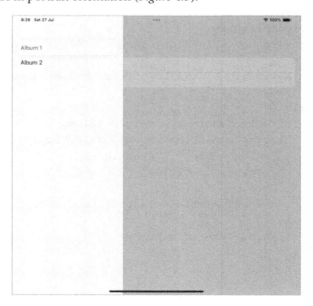

Figure 4.3: Two columns in SplitView in portrait orientation

Figures 4.2 and *4.3* show how our code runs on an iPad in portrait and landscape orientations. In portrait orientation, the sidebar view shows up in a drawer, and in landscape orientation, the screen is split, and both views are visible.

But what happens on an iPhone? Do we need to create a dedicated view for smaller devices? Let's see what happens with the same code on an iPhone (*Figure 4.4*):

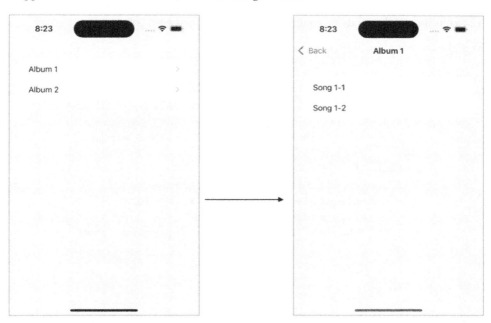

Figure 4.4: NavigationSplitView on an iPhone

Figure 4.4 shows that the same `NavigationSplitView` just works when running on an iPhone. On small devices, the `NavigationSplitView` constructs a one-page navigation mechanism, similar to what is seen in `NavigationStack` or even in UIKit's `UINavigationController`.

Now, let's make things a little bit more complex and add a third column.

Moving to three columns

In many apps, the data hierarchy is based on two levels. In our example, it is albums and songs, but in other cases, we can find groups and users, teams and players, or projects and tasks. Based on that, we will have to work with a three-level navigation system:

- Level 1: List of the first level items
- Level 2: Based on the first-level selection, the list of second-level items

- Level 3: Details of the selected second-level item

Even though we can present the details of the selected first-level item in a modal screen, we can consider showing it in a third column on an iPad screen.

In the *Creating a NavigationSplitView* section, we said that the `Detail` column shows information about the last selected item. This means that if we want to add another column, it will be between the `Detail` column and the `Sidebar` column – this is the `Content` column.

So, let's add a `Content` column to our music app:

```
var body: some View {
    NavigationSplitView {
        List(albums, selection: $selectedAlbum) { album
          in
            NavigationLink(album.title, value: album)
        }
    } content: {
        if let selectedAlbum = selectedAlbum {
            List(selectedAlbum.songs, selection:
              $selectedSong) { song in
                NavigationLink(song.title, value: song)
            }
            .navigationTitle(selectedAlbum.title)
        } else {
            Text("Select an album")
        }
    } detail: {
        if let selectedSong = selectedSong {
            VStack {
                Text("Song Title:
                  \(selectedSong.title)")
                Text("Artist: \(selectedSong.artist)")
            }
            .padding()
            .navigationTitle(selectedSong.title)
        } else {
            Text("Select a song")
        }
    }
}
```

In the preceding code example, we put the list of songs in our new `Content` block and the song details in the `Detail` column. We also used the same technique of `selectedSong` state variable and updated our UI accordingly.

Let's see how it looks now on an iPad (*Figure 4.5*):

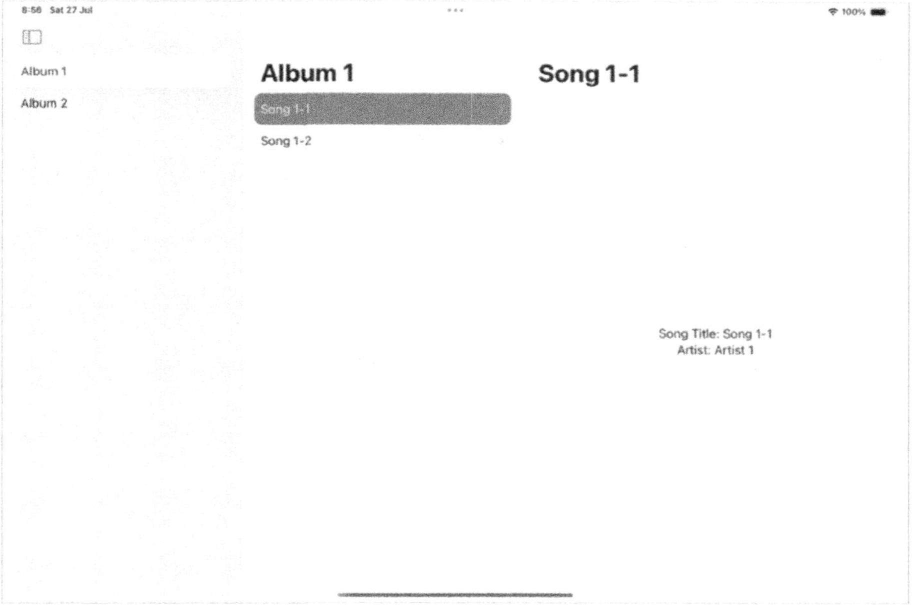

Figure 4.5: Three-columns NavigationSplitView on an iPad

Figure 4.5 shows a three-column `NavigationSplitView` on an iPad, now with the `Content` column showing the list of albums in the middle.

Summary

This chapter touches on a crucial topic in mobile development. Navigation has always been an issue, also in UIKit. However, we can achieve an effective navigation mechanism with thoughtful planning based on the product requirements and a balanced approach to flexibility and simplicity.

In this chapter, we went over the reasons why SwiftUI is a challenge, explored `NavigationStack`, reviewed the Coordinator pattern, and even discussed a column-based navigation with `NavigationSplitView`.

By now, we are more than capable of creating an amazing navigation in our app!

Our next chapter discusses something completely different but exciting: how to break our app's borders and add features outside our sandbox with WidgetKit.

5
Enhancing iOS Applications with WidgetKit

As iPhones have evolved over the years, new capabilities have been added to take advantage of the big screen, the memory capacity, and the powerful processor.

One of those capabilities is the home screen widgets – a great way to extend our apps and provide information and even interaction in new places.

In this chapter, we will cover the following topics:

- The idea of widgets
- Understanding how widgets work
- Add our first widget and build a timeline of entries
- Add a user-configurable widget
- Ensure our widgets are up to date
- Customize the widget animations
- Add user interactions such as buttons and toggles
- Add a control widget to the control center and lock screen

So, let's start with the basics – what is the idea of widgets?

Technical requirements

For this chapter, it's essential to download Xcode version 15.0 or higher from the App Store.

Ensure you're operating on the most recent version of macOS (Ventura or newer). Just search for Xcode in the App Store, choose the latest version, and proceed with the download. Open Xcode and complete any further setup instructions that appear. After Xcode is completely up and running, you can begin.

To gain additional capabilities, such as sharing data between the widget and the app, you must set up AppGroups and define your AppGroups in your profile.

Download the sample code from the following GitHub link:

`https://github.com/PacktPublishing/Mastering-iOS-18-Development/tree/main/Chapter%205`

The idea of widgets

Adding a widget is not a new concept in iOS or, in fact, in the Apple ecosystem.

Widgets existed long ago in 2005 in the *Tiger* version of macOS, as part of the Dashboard feature. Apple took that idea and introduced *Today Widgets* in the *Notification Center* in iOS 8, and in iOS 14, Apple introduced the home screen widgets, similar to the widgets that already exist in the Android operating system. In iOS 18, Apple added the ability for third-party applications to add widgets to the control center and the home screen.

The idea of widgets is not to act as a full-blown application – widgets are not supposed to be a mini-version of our app or one of its screens, but rather an extension of our current app's capabilities.

Widgets exist to enhance user convenience and productivity and, in general, the overall experience.

There are three key roles for widgets in iOS:

- **Information at a glance** – Widgets provide up-to-date and important information to the user about our app. It can be a delivery status, stock values, event calendars, or any other information that is useful on a day-to-day basis.
- **A shortcut to our app** – Tapping on a widget opens our app, and in many cases, a specific screen of our app. Opening our app using widgets is even more important in watchOS, where, unlike iOS, the springboard is not the user's default view. For many app developers, it's a great way of promoting their app and *fighting* for the user's attention on the home screen.
- **Performing basic actions** – Starting iOS 17, Apple added interactive widgets, allowing users to perform basic actions without opening their app, such as completing a task, opening the garage door, or accepting a payment request. In iOS 18, this capability went even further, and it's possible to add our widgets to the control center, or open them using the action button on iPhone 15 devices.

Going over the different Apple platforms, we can see that the idea of showing information at a glance is widespread – we've got home and lock screen widgets, complications, and live activities in iOS, padOS, macOS, and watchOS.

For example, the Yahoo! Weather app shows the weather in the user's current location and Apple's Reminders app shows the user's uncompleted reminders.

It's only natural for Apple to straighten the line between the different platforms into a single framework – *WidgetKit*.

Understanding how widgets work

As mentioned at the beginning of this chapter, widgets are not mini applications. Instead, widgets are simple views that show relevant information and are updated according to a declared timeline or app events.

Widgets run on a different process than the app. They receive a runtime to perform any code, so they work as static views, showing pre-made information to our users. But, since our user's data is being constantly updated, we can create an array of entries, each with information and a date. The *WidgetCenter* is responsible for creating a different view for each one of the entries, storing it, and replacing the widget UI according to the entries' dates. This array of entries is called a **timeline**.

One good example is the *Next Event* widget. The *Next Event* widget shows the next event in our calendar, and since we have access to our user's calendar, we can build a timeline and refresh the widget data based on the calendar event's list. All we need to provide is the timeline including the different data for each timeline entry.

Using a timeline to update the widget's content makes the widget an extremely effective way to present information to the user, both in battery usage and processing time.

However, the timeline also produces some challenges in the way we work with widgets because, unlike the *Next Event* widget, not every timeline can be built up front.

But let's wait before we dive into the solution to our problems and try to add our first widget.

Adding a widget

Widgets operate and live outside of our app, therefor they are considered to be an *extension* of our app.

To add a new widget – we need to a new **Widget Extension** target by selecting **File** -> **New** -> **Target**....

Then, in the **Choose a template for your new target** window, we search for a widget and add the widget extension (see *Figure 5.1*):

Figure 5.1: The Choose a template for your new target window

After clicking on **Next**, we should provide a name for our widget, just like any target we add. In addition, uncheck the *Include Configuration App Intent* checkbox.

Once the widget is added, we can see a new target with the name we provided. Xcode creates a few files for us as part of the widget template (assuming that the target name is `MyWidget`):

- `MyWidgetBundle` – The widget bundle is a container for the different widgets our extension holds. Currently, we have only one widget, but it is possible to add more.
- `MyWidget` – Contains the widget code itself, including its UI and configuration.
- `Assets` – An asset catalog specifically for the widget extension.
- `Info.plist` – Just like any target, the widget extension contains a `plist` file with general information about the widget extension.

Now, it's time to clarify what a widget is – the fact that we have different sizes for a widget doesn't mean they are different widgets, as the same widget can have multiple sizes. A different widget is usually a different product, a different UI, and a different use case. In our case, the widget bundle describes the different widgets and not the different widget sizes.

Now that we have added a widget to our project, we can run our app and add the new widget to our springboard (*Figure 5.2*):

Figure 5.2: Our new template widget in the springboard

We can see in *Figure 5.2* that the new widget consists of the current time and some emoji. This is a good time to play with it and to try adding additional widget sizes.

Configuring our widget

The way we set up our widget's look and behavior is by determining its configuration. We have several configurations to work with, and they all conform to a protocol named `WidgetConfiguration`.

One of the configurations available for us is `StaticConfiguration`. `StaticConfiguration` allows us to create a widget that has no user-configurable options.

Let's have a look at the `StaticConfiguration` that Xcode provides when we add a new widget:

```
struct MyWidget: Widget {
    let kind: String = "MyWidget"

    var body: some WidgetConfiguration {
        StaticConfiguration(kind: kind, provider:
```

```
            Provider()) { entry in
                MyWidgetEntryView(entry: entry)
                    .containerBackground(.fill.tertiary, for:
                        .widget)
            }
            .configurationDisplayName("My Widget")
            .description("This is an example widget.")
        }
    }
```

We can see that `StaticConfiguration` has several properties shared with all configuration types. Let's see them in depth, here:

- `kind` – This is the widget configuration unique identifier. It helps us send requests to a specific widget configuration using the *WidgetCenter*.
- `configurationDisplayName` – This is the widget display name as it appears for the user when he wants to pick the right widget to add.
- `description` – This is the widget's description that is shown to the user, next to its display name.

Besides these three parameters, we have additional important parameters. `supportedFamilies` determines the different sizes the widget supports. Here's an example of how to limit the widget to appear only in medium size:

```
.supportedFamilies([.systemMedium])
```

Another property is `backgroundTask`, which allows our widget to perform a background operation when the system gives it time.

Notice that `WidgetConfiguration` is just a protocol – when creating a widget, we need to return, in the widget body, a structure that conforms to that protocol, and `StaticConfiguration` is just one way to do that.

Currently, there are three configurations available for us:

- `StaticConfiguration` – As mentioned earlier, this configuration allows us to create a non-user configurable widget
- `AppIntentConfiguration` – This enables the user to customize their widget, for example, selecting a city for a weather widget, or a specific list for the reminders app
- `ActivityConfiguration` – This configuration shows live data for the Live Activity widget

A widget can contain only one configuration. If we need to have more than one configuration, that's a good sign we need to create several widgets with different configurations and share some of our code between them.

All these widget configurations sound exciting! Let's start exploring them by starting with the `StaticConfiguration`.

Working with static configuration

A static widget is a widget that has no user-configurable options. For example, a widget that shows the current time in a specific city cannot be static because the user needs to specify a city or a location for the widget.

However, a good example of a static widget is a calendar widget that shows a view of the whole month and marks the current day, or a music app widget that shows the songs that have been played recently.

Even though both the calendar and the music app widgets show information not updated by the user, they need to update themselves every once in a while.

If we look back at the static configuration example (in the *Configuring our widget* section), we can see a parameter called `provider`, which contains a parameter for the view builder closure named `entry`.

Using `provider` and `entry`, we can provide data to our widget across time in an efficient way.

One key aspect of Widgets is providing data over time, and we do that using the Timeline provider. Now, let's understand what Timeline Provider means.

Understanding the Timeline Provider for Widgets

There's a reason why it took Apple almost 14 years to support widgets on the iOS home screen. The primary reason is performance, both power and memory performance. While today's devices are highly capable, having numerous active widgets on the *Springboard* can consume a significant amount of power. Hence, we need to find more efficient ways to load our widgets efficiently.

We mentioned efficiency in the *Understanding how widgets work* section, so let's get down to the details. Unlike apps, widgets are not active even when they are visible. We can "wake" these widgets at specific times to reload their views. To set the specific periods, we need to create a timeline – an array of entries that contain points in time and relevant data.

For example, if we want to reload a calendar widget that displays the next event, we can create a timeline that holds an array of entries, one for each event. Each entry holds an event time and the name of the event that comes afterward.

Conversely, if we want a calendar widget that displays full-day information, we may want to create a timeline with an entry for each day. In this case, each entry holds the time of the beginning of the day and the list of events that day.

Creating a longer timeline can maximize the frequency of updates for our widget.

Now, let's turn to code and create our first timeline. Here is an example of a timeline provider that displays the next event:

```
struct EventEntry: TimelineEntry {
    let date: Date
    let nextEvent: String
}

struct Provider: TimelineProvider {
    func placeholder(in context: Context) -> EventEntry {
        EventEntry(date: Date(), nextEvent: "Loading")
    }

    func getSnapshot(in context: Context, completion:
      @escaping (EventEntry) -> Void) {
        let entry = EventEntry(date: Date(), nextEvent: "Go
          to the book store")
        completion(entry)
    }

    func getTimeline(in context: Context, completion:
      @escaping (Timeline<EventEntry>) -> Void) {
        let entries: [EventEntry] = getListOfEnties()
        let timeline = Timeline(entries: entries, policy:
          .atEnd)
        completion(timeline)
    }

    func getListOfEnties()->[EventEntry] {
        ...
    }
}
```

The preceding code consists of two structs – `EventEntry` and `Provider`.

`EventEntry` is a struct that conforms to `TimeLineEntry` protocol. The `TimeLineEntry` protocol represents a single entry in the widget timeline. The protocol contains a required variable named date:

```
var date: Date { get }
```

The `date` variable contains the entry point in time where we expect our widget to reload. Other than `date`, we added another variable that represents the entry's next event title named `nextEvent`.

Our second struct is `Provider`. The `Provider` struct conforms to `TimeLineProvider`. The goal of the `Provider` struct is to generate a timeline so the *WidgetCenter* can reload the widget when needed. Let's see how the `Provider` does that.

Generating a timeline

I mentioned earlier that a timeline is an array of timeline entries, but the reality is a little bit more complex than that. Looking at the timeline provider implementation, we can see several functions that help us to deliver a static UI at any given time.

> **The Provider struct is a protocol implementation**
>
> There's no need to call the `Provider` functions directly. We pass the timeline provider to the widget configuration, and the configuration uses the `Provider` functions when needed.

The first and primary function is `getTimeLine`. Let's look at the implementation of the `getTimeline` function here:

```
func getTimeline(in context: Context, completion: @escaping
    (Timeline<EventEntry>) -> Void) {
        let entries: [EventEntry] = getListOfEntries()
        let timeline = Timeline(entries: entries, policy:
          .atEnd)
        completion(timeline)
}
```

The `getTimeline()` function creates an array of entries, wraps them in a `Timeline` struct, and returns it using the `completion` closure. There are two interesting things we can see here – the `Context` parameter and the **Timeline reload policy**:

- `Context` – The `Context` parameter contains information about the widget environment, such as the widget family (is it a small widget? Perhaps medium?), or the actual widget size. If the widget UI shows more information when it is large, we probably want to load more data into our timeline entry. But the most important information here is probably the `isPreview` property, which indicates whether the widget appears in the widget gallery. Generally speaking, it is best practice to show real user data in our widget in the widget gallery, but that's not only possible due to security or networking issues. Therefore, we can provide mock data for the widget gallery by checking the `isPreview` property.
- `policy` – The timeline we provide to our widget has a final number of entries. So, what happens when they are done, and the timeline reaches its end? That's exactly the role of the `policy` parameter when it describes the timeline reload behavior. There are several options – `atEnd` (*WidgetKit* requests a new timeline), `never` (*WidgetKit* doesn't ask for a new timeline), and

after(date:Date) (*WidgetKit* generates a new timeline in a specific date). The policy helps the *WidgetCenter* to optimize the timeline reloading mechanism better.

Before we continue, a few words about timeline reloading optimization. The fact that we want to build our timeline as long as possible doesn't mean that our widget needs to constantly reload. The *WidgetCenter* has a "budget" for each widget on the home screen, specifying times during the day when it performs refreshes. It's in our interest to optimize the way our timeline is structured and to "save" the system budget. Carefully planning the timeline entries and reload policy can help us achieve relevant, event-driven refresh intervals.

Going back to the `TimelineProvider` protocol, we can see additional two functions – `placeholder` and `getSnapshot`. Let's implement them.

The first function is `getTimeline`, which returns a `Timeline` structure containing a list of entries with actual data for specific periods. But is it enough for our widget to be fully functional?

The answer is no – there are two more cases when providing actual data may not be sufficient.

The `placeholder` function answers the first use case. When the user adds a widget to their home screen, *WidgetKit* needs to display something immediately, before the widget fetches or updates real data from our app. The `placeholder` function returns temporary data just to show something to the user:

```
func placeholder(in context: Context) -> EventEntry {
    EventEntry(date: Date(), nextEvent: "English
      class")
}
```

In our example, we can see a `placeholder` function that returns the `English class` text.

It is important to return temporary data instead of a loading indicator, for example, and that's because we want our user experience to be consistent and smooth. It is also better to be creative and come up with elegant information. For example, if our widget has a timer or a time, it is a good idea to show 00:00 to indicate to the user that a timer should appear here.

The second function is `getSnapshot`. The `getSnapshot` function is even more important than `placeholder`. When the user browses the widget gallery, the system presents the different widgets. These widgets are being presented without the system-generated timelines.

The `getSnapshot` function returns a `TimelineEntry`-based struct with data to present in the widget gallery.

Here's an example of a `getSnapshot` function:

```
func getSnapshot(in context: Context, completion:
  @escaping (EventEntry) -> Void) {
    let entry = EventEntry(date: Date(), nextEvent: "Go
```

```
            to the book store")
        completion(entry)
    }
```

In this code, the `getSnapshot` function returns an example event with the current date. This snapshot demonstrates to the user the purpose of our widget easily.

Note that in both `placeholder` and `getSnapshot`, we have the same `Context` parameter as the one we had in the `getTimeline` function. We need the `Context` for the same reason as before – to understand the environment surrounding our widget.

Now that we understand how to generate a timeline provider, let's discuss the `TimelineEntry`.

Building our TimelineEntry structure

We can see by now that the `TimelineProvider` protocol is straightforward as there are only three functions to implement. One of the things we need to design here is `TimelineEntry`, and the reason for its importance is that it holds the information we need not only to determine when to present information but also what to present.

The structure of `TimelineEntry` needs to fit our widget goal and be aligned with its UI. Because we pre-generate all the entries according to a timeline, we should perform all the calculations in advance and generate a structure that can help update the widget content easily.

In fact, `TimelineEntry` may consist of four components:

- `date` – The date when we want our widget to reload the specific entry information. Notice that in most cases, the `date` property is not part of the information presented on the screen. For example, in a calendar widget, we probably have a date property as part of the `TimelineEntry` protocol, and something like `eventDate` for the actual event time.
- **Information to display** – It is better to add properties that will make rendering our widget easier later on. For example, adding properties such as `title`, `bodyText`, and `timeString`, can simplify our code and even increase performance.
- **Metadata** – If we want to support some interaction with the widget, we need to hold some metadata related to the widget data model. For example, a calendar widget might hold some metadata containing the event ID, a contacts widget might have metadata containing the contact ID, and so on. Remember that once the widget is presented, `TimelineEntry` is all that we have when the user interacts with it.
- **Relevance** – The `relevance` property is an optional property that we have as part of the `TimelineEntry` protocol. In the `relevance` property, we can determine the relevance priority of the entry to the user. For example, a to-do app that shows the next task to the user may want to set a high score to an entry with a critical task. Or, a sports app that shows the latest news in a widget may want to set a high score for entries that contain news about the

user's favorite team. The entries' relevance values help *WidgetKit* to decide how and when to present widgets in the system. For example, *WidgetKit* may decide to rotate the stack widget and show a widget with high-relevance information. Let's see an example how to set `relevance` for a `TimeLineEntry`:

```
struct EventEntry: TimelineEntry {
    let date: Date
    let nextEvent: String
    var relevance: TimelineEntryRelevance?
}
let entry = EventEntry(date: date, nextEvent: "Go to
  the book store", relevance:
  TimelineEntryRelevance(score: 1.0))
```

In this code, we added the `relevance` property to our `EventEntry` struct and set a score of `1.0`. It is worth noting that any efforts to manipulate the system and set high scores for all entries won't succeed – Apple has built an algorithm that filters out widgets that have unrealistic values. As with many iOS frameworks, this is a situation where we need to follow the platform's intended usage guidelines.

Now that we have created a timeline, let's turn to the main topic, which is building our widget UI.

Building our widget UI

Creating a timeline of entries is critical for our widget to provide accurate and relevant information to the user. But to do that, we also need to render the widget UI. The place where we do that is in the widget's structure, as we saw at the beginning of this chapter in the *Configuring our widget* section.

Let's see the configuration again:

```
StaticConfiguration(kind: kind, provider: Provider()) {
    entry in
        MyWidgetEntryView(entry: entry)
            .containerBackground(.fill.tertiary, for:
            .widget)
}
```

As we can see, the `StaticConfiguration` has a view builder that returns a SwiftUI view, and this is probably the first thing we need to understand in *WidgetKit* – widgets are built with SwiftUI only. If you still haven't got any experience with SwiftUI, *WidgetKit* is a great opportunity to start.

Something that might have caught your attention is the `containerBackground` view modifier. If you remember, we have discussed how widgets now can be shown in different places in the Apple ecosystem – *iOS* (both home screen and lock screen), *padOS*, *macOS*, and *watchOS*. But the primary issue with having our widgets on different platforms might be the widget's background.

Adding the `containerBackground` view modifier ensures that the widget's background adjusts itself to its container and always looks good, no matter where it appears.

If we look at our code example again, we can see that `MyWidgetEntryView` receives one parameter, which is the current timeline entry. Let's see what we can learn from it.

Working with timeline entries

Connecting the timeline entry to the widget view is the core of how widgets work. The main role of *WidgetCenter* is to generate a timeline and provide our widget with the right timeline entry at the right time.

The widget configuration view builder has one parameter, a specific timeline entry, so we can return a widget view with relevant data.

Here's an example of a widget view that uses a specific timeline entry:

```
struct MyWidgetEntryView: View {
    let entry: EventEntry

    var body: some View {
        VStack(alignment: .leading) {
            Text("Next Event:")
                .font(.headline)
            Text(entry.nextEventTitle)
                .font(.title)
                .foregroundColor(.blue)
            Text("Time: \(entry.nextEventTime)")
                .font(.subheadline)
            Spacer()
        }
        .padding()
    }
}
```

This code example shows a simple view that shows the next event title and time while using the timeline entry.

There are two things we can learn from the way the timeline entry works with the widget view:

- **The entry should contain all the widget's data** – We discussed it when we talked about the timeline provider, but now we can see why. Widgets need to be as static and simple as possible. We don't want to perform any data fetching operations while the view is displayed.
- **There is no state** – Unlike regular SwiftUI views, our widget view doesn't have a state. There are cases where we probably would want to see different views for different situations. For example,

in our *next event* widget example, maybe we want to show a *connect to your calendar* message if the user hasn't approved his calendar permissions. To do that, we need to generate different timeline entries and perhaps show a different view in the static configuration closure. Either way, we should do these checks in advance.

Even though the widgets are naturally static, their UI doesn't have to stay static and bold. In `WidgetKit`, it is possible to bring life to our widget by animating the changes.

Adding animations

We already know how animations in iOS development work – view animations work by transitioning between two or more states. For example, if a specific view has an opacity of `1.0` and we change it to `0.5`, UIKit and SwiftUI can animate that change if we like.

Widgets are written in SwiftUI, and in SwiftUI, we can animate state changes. However, widgets don't use state at all. Instead, we change the widget content using the timeline provider (perhaps we can say that, in a way, the timeline entry is our widget state).

Starting with iOS 16, whenever the *WidgetCenter* reloads a widget and changes its content using a new entry, it performs this transition automatically.

Can we customize this animation even if we don't have a state in widgets? Of course we can, using `contentTransition`.

As mentioned, in most cases, SwiftUI performs animations based on a state change. For example, look at the following code:

```
@State private var isRed = false

var body: some View {
    VStack {
        Color(isRed ? .red : .blue)
            .frame(width: 100, height: 100)
            .cornerRadius(10)
        
        Button("Change Color") {
            withAnimation {
                self.isRed.toggle()
            }
        }
    }
}
```

In this code example, we have a view and a button. Tapping on the button changes the view color, and it does that using the `withAnimation` function. Clearly, that can't work in a widget because we need a state to do that.

Instead, what we need to do is define how the content changes when it's animated. To do that, we can use `contentTransition`:

```
Color(isRed ? .red : .blue)
    .frame(width: 100, height: 100)
    .cornerRadius(10)
    .contentTransition(.opacity)

Button("Change Color") {
    withAnimation() {
        self.isRed.toggle()
    }
}
```

`contentTransition` is a view modifier we can add to views to define their transition method. Imagine that all content changes in widgets are done with `withAnimation` in mind and all we have to do is to change the transition method.

Take, for example, the following code snippet:

```
Text(text).contentTransition(.numericText())
```

When changing the text using the `withAnimation()` function, it will change its content with a nice numeric transition (you can try it yourself). If you are not familiar with the `withAnimation` function, *Chapter 6* provides a brief discussion on it.

In widgets, all we need to do is to add these to views with content that is based on our timeline entry, and SwiftUI will take care of the animation itself.

Look at our widget again, now with `contentTransition`:

```
struct MyWidgetEntryView : View {
    var entry: Provider.Entry
    
    var body: some View {
        VStack {
            Text("Time:")
            Text(entry.nextEventTime, style: .time)
```

```
                Text("Next Event")
                Text(entry.nextEvent)
                    .contentTransition(.numericText())
        }
    }
}
```

Even though there is no state or `withAnimation` function, the `nextEvent` title will animate its transition. The `contentTransiton` view modifier has additional options, such as opacity and symbol effects. Despite the fact that it is not designed explicitly for widgets, it's the best way to make our widgets more alive.

Customize our widget

Up until now, we have discussed widgets based on a `staticConfiguration`. The `staticConfiguration` set is great for most widgets. However, there are cases where we want to provide our users the ability to customize and configure their widgets with additional entities.

Going back to our calendar widget, we want to allow the user to filter the next event information based on a specific calendar.

To do that, we'll start by creating a new file and add a struct called `CalendarWidgetIntent` that conforms to `WidgetConfigurationIntent`.

Adding intent

A `WidgetConfigurationIntent` is an App Intent we can use to configure widgets, and our `CalendarWidgetIntent` contains all the configuration information we need.

Here is a basic `CalendarWidgetIntent` implementation:

```
struct CalendarWidgetIntent: WidgetConfigurationIntent {
    static var title: LocalizedStringResource = "Select
      Calendar"

    @Parameter(title: "Calendar") var calendar:
      CalendarEntity
}
```

In the preceding code, we can see two properties:

- `title` – The title of the intent. It is important to note that we don't see the title in the widget configuration string but rather in Siri Shortcuts. But we must add it since it is part of the `AppIntent` protocol (the `WidgetConfigurationIntent` inheritance from `AppIntent` protocol).

- `calendar`– This is the widget parameter that allows the user to configure the calendar the event belongs to. We can see that the `calendar` variable is prefixed by the `@Parameter` macro, which manages this property for the user's configuration.

Now, let's add the App Intent.

Adding AppEntity

As you have noticed, the calendar variable is based on a type called `CalendarEntity`.

If we want to support our own entity type, it needs to conform to `AppEntity`. Let's see the `CalendarEntity` type implementation:

```
struct CalendarEntity: AppEntity {
    let id: String
    let name: String
    
    static var typeDisplayRepresentation:
       TypeDisplayRepresentation = "Calendar"
    static var defaultQuery = CalendarQuery()
    
    var displayRepresentation: DisplayRepresentation {
        DisplayRepresentation(title: name)
    }
}
```

The `CalendarEntity` struct represents the data model for the `intent` parameter. First, we need to add the parameters we need in order to support the item when displaying the widget, such as `id` and `name`. Next, we'll add some representation variables, such as `typeDisplayRepresentation` and `displayRepresentation`.

Finally, we'll add a static variable that handles the actual data fetching, and that's the `defaultQuery` property. Remember that the user needs to select the desired calendar based on a list of calendars. To do that, we need to provide *WidgetKit* with a way to query our data to support the selection UI flow.

So, what does the query look like? Let's find out.

Building the EntityQuery

Sometimes, having a list of options for the user relies on a data store, and sometimes on static information.

Regardless of the model type, if we want to provide options to the user, we need to have a simple and effective interface to work with, and that's what the `EntityQuery` protocol is for.

In our current `AppIntent` example, we let the user choose one of its calendars, so we need to build a struct named `CalendarQuery` that conforms to `EntityQuery`.

Let's look at a simple `CalendarQuery` example:

```
struct CalendarQuery: EntityQuery {

    func entities(for identifiers: [CalendarEntity.ID])
      async throws -> [CalendarEntity] {
        allCalendars.filter { identifiers.contains($0.id) }
    }

    func suggestedEntities() async throws ->
      [CalendarEntity] {
        allCalendars
    }

    func defaultResult() async -> CalendarEntity? {
        nil
    }
}
```

Assume that `allCalendars` is an array containing all the user calendars.

In this case, `CalendarQuery` implements three methods. Let's quickly go over them:

- `entities(for identifiers:)` – This function returns calendar entities based on a list of IDs. *WidgetKit* uses it to show the selected calendar
- `suggestedEntities()` – This returns the list of entities in the pop-up menu
- `defaultResult()` – This returns the value when nothing is selected

Now, let's see how it looks (*Figure 5.3*):

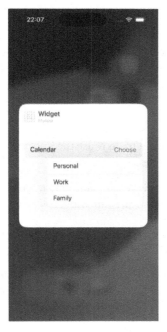

Figure 5.3: The widget configuration menu

In *Figure 5.3*, we can see the **Widget** configuration screen with the parameter we declared (**Calendar**) and the calendar names in the pop-up menu. It is worth mentioning that we can define different types of parameters, such as `Bool` or `String`, and *WidgetKit* will create their corresponding input control.

Let's flip to the other side now and go to the widget UI to use the `AppEntity` the user selected.

Using the AppEntity in our Widget

Going back to our widget code, let's examine the widget configuration code again:

```
AppIntentConfiguration(kind: kind, intent:
    CalendarWidgetIntent.self, provider:
    ConfigurableProvider(), content: { entry in
        ConfigurableWidgetView(entry: entry)
    })
```

The `AppIntentConfiguration` struct has an important property, which is the intent type it uses, and in this case, it is `CalendarWidgetIntent`. If we go back to the *Customize our widget* section, we can see that `CalendarWidgetIntent` contains all the information we need to present our widget according to the user configuration.

Indeed, the timeline provider is now conforming to a different protocol, `AppIntentTimelineProvider`, which supports the intent configuration now. Let's see how it creates a timeline:

```
struct ConfigurableProvider: AppIntentTimelineProvider {
    func timeline(for configuration: CalendarWidgetIntent,
        in context: Context) async ->
            Timeline<ConfiguredNextEventEntry>
```

We can see that the timeline function inside `ConfigurableProvider` now receives the configuration parameter. From this point, all we need to do is use the information we have inside the configuration and create the relevant timeline entries.

By now, we know how to set up a new widget, animate it, create its timeline, and even let the user configure it. Next, we'll learn how to ensure our widgets stay up to date.

Keeping our widgets up to date

We have learned that we need to look ahead and create a timeline with different entries and dates to keep our widget up to date. But how does our widget work under the hood?

Widgets don't get any running time – once we generate the timeline entries, *WidgetCenter* generates their different views, keeps them persistently, and just switches them according to the provided timeline.

So, there's no way to update our widget without reloading the timeline, and when we created our timeline, we had to define its reload policy:

```
let timeline = Timeline(entries: entries, policy: .atEnd)
```

However, sometimes, we want to instruct `WidgetCenter` to reload the timeline immediately, due to data changes or any other alterations.

Let's see how it happens.

Reload widgets using the WidgetCenter

Throughout the chapter, I have mentioned *WidgetCenter* frequently but I haven't explained what it means.

WidgetCenter is an object that holds information about the different configured widgets currently used, and it also provides an option to reload them.

To use `WidgetCenter`, we need to call the `shared` property to access its singleton reference:

```
WidgetCenter.shared
```

The difference between *WidgetCenter* and the rest of the code we have handled up until now is the fact that we call *WidgetCenter* from the app and not the widget extension.

Let's see how we can call the `WidgetCenter` to get a list of active widgets:

```
func getConfigurations() {
    WidgetCenter.shared.getCurrentConfigurations { result
      in
        if let widgets = try? result.get() {
            // handle our widgets
        }
    }
}
```

The `getCurrentConfigurations` function uses a closure to return an array of active widgets. Each one of them is the `WidgetInfo` type – a structure that contains information about a specific configured widget.

The `WidgetInfo` structure has three properties – kind, family, and configuration:

- `kind` – This is the string we set when we created the widget configuration (look again at the *Configuring our widget* section).
- `family` – The family size of the widget – small, medium, or large.
- `configuration` – The intent that contains user configuration information. The `configuration` property is optional.

If needed, we can use that information to reload the timeline of a specific kind of widget. For example, if we want to reload widgets with the kind of `MyWidget`, we need to call the following:

`WidgetCenter.shared.reloadTimelines(ofKind: "MyWidget")`

Notice that the function says `Timelines` and not `Timeline`, as it is possible to have several widgets of the same kind.

If we want to reload all our app widgets, we can call the `reloadAllTimelines()` function:

`WidgetCenter.shared.reloadAllTimelines()`

There are several great use cases for reloading our widget timeline, such as when we get a push notification, or when the user data or settings have changed. If you remember, when we discussed the widget timeline in the *Generating a timeline* section, we talked about the fact that widgets have a certain budget for the amount of reloading they can do each day. But the good news is that calling the `reloadTimelines` or `reloadAllTimelines` functions doesn't count in this budget if our app is in the foreground or uses some other technique, such as playing audio in the background.

In most cases, `reloadTimelines` works well when the updated data is already on the device or in our app. But what should we do when the local persistent store is not updated?

We perform a network request, of course!

Go to the network for updates

Performing a network request to update local data is a typical operation in mobile apps. But how does it work in widgets?

Let's look at the `getTimeline` function again:

```
func getTimeline(in context: Context, completion: @escaping
(Timeline<Entry>) -> ())
```

We can see that the `getTimeline` function is an asynchronous function. It means that when we build our timeline, we can perform async operations such as open URL sessions and fetching data.

Let's see an example of requesting the next calendar events:

```
func getTimeline(in context: Context, completion: @escaping
    (Timeline<SimpleEntry>) -> Void) {
        var entries: [SimpleEntry] = []
        calendarService.fetchNextEvents { result in
            switch result {
            case .success(let events):
                for event in events {
                    let entry = SimpleEntry(date:
                        event.alertTime, nextEvent:
                            event.title, nextEventTime:
                                event.date)
                    entries.append(entry)
                }
            case .failure(let error):
                print("Error fetching next events:
                    \(error.localizedDescription)")
            }
            let timeline = Timeline(entries: entries, policy:
                .atEnd)
            completion(timeline)
        }
    }
```

The `getTimeline` function implementation is similar to the previous `getTimeline` implementation we saw in the *Generating a timeline* section, and this time, we are fetching the events using the `calendarService` instance. The `calendarService` goes to our server and returns an array of events. Afterward, we loop the events, generate timeline entries, and return a timeline using the `completion` block.

Up until now, we have seen how to create a widget, animate it, and ensure it is updated as much as we can. But if we want to make our widget shine, we need to add some user-interactive capabilities.

Interacting with our widget

Besides providing us with a glance at our app information, widgets are a great way to open our app in a specific location or manipulate data without even opening the app.

As mobile developers, we sometimes wonder why implementing user interaction with a widget is such a big deal. After all, our users interact with our app daily, so why is it such a problem? But when we remember that widgets don't really run, we can understand the challenge.

The most basic way we have to allow interaction with our widgets is by using deep links.

Opening a specific screen using links

If you are not familiar with the concept of deep links, now is the time to straighten things out. A deep link is a link that opens our app on a specific screen. Today's deep links format is similar to website URLs. For example, a deep link that opens our app in a specific calendar event screen can look something like this:

```
http://www.myGreatCalendarAp.com/event/<eventID>/
```

To do that, the app needs to do three things:

- *Register to that specific domain* by placing a special JSON file on the relevant server
- Add the domain *entitlement* to our app
- Respond to *launching the app from a deep link*, parse the URL, and direct the user to the corresponding location within the app

To learn more about deep links, I recommend reading about it in the Apple Developer website:

```
https://developer.apple.com/documentation/xcode/allowing-apps-and-
websites-to-link-to-your-content
```

Going back to our topic, let's see an example of adding a deep link to our Next Event widget:

```
struct MyWidgetEntryView : View {
    var entry: Provider.Entry

    var body: some View {
        VStack {
            Text("Time:")
            Text(entry.nextEventTime, style: .time)

            Text("Next Event")
            Text(entry.nextEvent)
                .widgetURL(URL(string:
 "https://www.myGreatCalendarApp.com/event/\(entry.eventID)/"))
        }
    }
}
```

In the preceding code example, we can see that we added a view modifier called `widgetURL` to the Next Event Text component.

The Next Event Text component is indeed the view that accepts the user's touch and opens the app in the specific deep link. But when the widget is small (`.systemSmall`), we can add only one deep link that is acceptable in the whole widget.

In widgets with medium and large sizes, we can add multiple links to multiple components.

It is worth noting that in terms of security, deep links work even when the device is locked, but require *FaceID* or passcode when tapping on them.

In iOS 17, deep links are not the only option we have to allow users to interact with our widget, as it is possible to add buttons and toggles as well.

Adding interactive capabilities

Deep links in widgets are great, but they have one big problem – tapping on the widget always opens the app. But, sometimes, we want to update data or confirm something without entering the app. For example, maybe we want to accept a calendar invitation, approve a payment, or mark a task as completed.

Because widgets don't actually run, it's a challenge to respond to a user interaction. Fortunately, there is a solution that we have already encountered in configurable widgets, and that's App Intents.

Using App Intents to add interactive widgets

App Intents made especially for this kind of use case – to allow runtime for specific actions.

So, how do App Intents help us? Let's look at *Figure 5.4*:

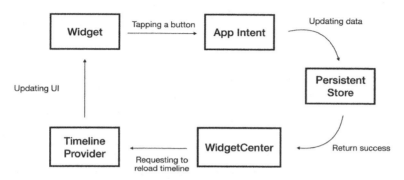

Figure 5.4: App Intent event flow

In *Figure 5.4*, we can see that the first stage is tapping on a button inside the widget. Starting from iOS 17, the *WidgetKit* framework has its own type of button, which can be linked to a specific App Intent:

```
Button("Turn \(entry.isAlarm ? "Off" : "On") Alarm" , role:
    nil, intent: MyWidgetIntent(eventID: entry.eventID))
```

The goal of the button here is to toggle the alarm for the event on and off. We can see that the title is changed according to the entry.isAlaram value. But what's more interesting here is that we have an additional parameter called Intent, where we pass a struct named MyWidgetIntent along with the eventID.

Let's talk about the App Intent, but this time, in the context of user interaction.

Performing the data change using the intent

We have already said that the widget doesn't manage any state. Therefore, the real widgets state is some kind of a combination between a local store and the timeline provider.

The MyWidgetIntent receives an eventID and is responsible for reaching out to EventKit and updating the actual event alarm information.

Let's look at the App Intent:

```
struct MyWidgetIntent: AppIntent {
    init() {
    
    }
```

```
        var eventID: String = ""

    init(eventID: String) {
        self.eventID = eventID
    }

    static var title: LocalizedStringResource = "Changing '
      event alarm settings."

    func perform() async throws -> some IntentResult {
        // working with EventKit and updating the event alarm data.
        return .result()
    }
}
```

Besides the `LocalizedStringResource` static property that we discussed in the *Customize our widget* section, we have one primary function called `perform()`. The `perform()` function executes when the user taps on the button that is linked to that App Intent. Notice that the `perform()` function is also an async function that lets us perform heavier tasks, such as writing to the database or even performing a URL request.

Once the `perform()` function completes its execution, the App Intent triggers the *WidgetCenter*.

Updating the widget UI

Now that the local store is updated, it's time for the *WidgetCenter* to reload the Timeline Provider. We should already be familiar with that process – the Timeline Provider fetches the relevant local data and builds a timeline based on the changes we just performed. At the end, the widget UI is being updated.

Working with App Intent is also great if we want to share code execution between different app components. For example, we can share logic code between our widget and the Siri Shortcut.

We should remember that even if the widget could have a runtime of its own, it is still a good practice to separate our code for better flexibility and modularity.

Another great usage for App Intent is the control widget, another great addition to iOS 18. Let's go over it now.

Adding a control widget

WidgetKit provides ways to present our apps in the springboard. However, it doesn't stop there. Starting iOS 18, it is possible to present widgets in the control center and on the lock screen and even attach an App Intent to the action button in iPhone 15 Pro.

Adding a widget to the control center or the lock screen is easy.

Similar to how we create a widget by conforming to the Widget protocol, we need to conform to the `ControlWidget` protocol to create a control widget. For example, imagine we have an app that helps us control smart home accessories, and we want to create a widget that opens and closes our home's main door. Let's start by creating a simple control widget called `MaindoorControl`:

```
struct MaindoorControl: ControlWidget {
    var body: some ControlWidgetConfiguration {
        StaticControlConfiguration(
            kind: "com.avitsadok.MaindoorControl"
        ) {

// rest of the widget goes here
        }
    }
}
```

In this code example, the `MaindoorControl` widget contains the body variable from the time of `ControlWidgetConfiguration`. This is very similar to how we created a home screen widget under the *Configuring our widget* section.

In this case, we return an instance of the `StaticControlConfiguration` type, which means we don't give the user the ability to configure it. However, similar to the home screen widget, we can also add a user-configurable control widget by returning `AppIntentControlConfiguration` (look in the *Customize our widget* section).

We can add two control widget controls – a toggle and a button. In the case of controlling our home's main door state, we need to add a toggle. Let's modify our code and add a `ControlWidgetToggle` instance:

```
struct MaindoorControl: ControlWidget {
    var body: some ControlWidgetConfiguration {
        StaticControlConfiguration(
            kind: "com.avitsadok.MaindoorControl"
        ) {
            ControlWidgetToggle(
                "Main door control",
                isOn: HouseManager.shared.isOpen,
                action: MaindoorIntent()
            ) { isOn in
                Label(isOn ? "Opened" : "Closed",
                      systemImage: isOn ?
                        "door.left.hand.open" :
                          "door.left.hand.closed")
            }
        }
```

            }
    }

In this code example, we add the `ControlWidgetToggle`, containing the following parameters:

- **A title** – The widget title that appears for the widget in the widget gallery.
- `isOn` – Here, we connect the widget to the actual state in our app.
- `action` – The App Intent that runs when the user taps our control widget. We'll cover that later in this section.
- **A view builder** – In the view builder, we define how a `Label` displays the control state's title and an image.

The widget instance is straightforward. Let's see how it looks in our control center (*Figure 5.5*):

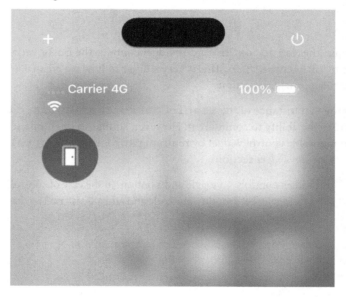

Figure 5.5: Our control widget in the control center

Figure 5.5 shows the control center's main door control widget. However, there's another side to our control widget – connecting the control widget to the action of opening and closing the main door. Let's look at the `MaindoorIntent` struct we saw in the `action` parameter:

```
struct MaindoorIntent: SetValueIntent {
    static let title: LocalizedStringResource = "Maindoor
       opening"

    @Parameter(title: "is open")
```

```
    var value: Bool

    func perform() throws -> some IntentResult {
      HouseManager.shared.isOpen = value
        return .result()
    }
}
```

In this code example, we see the `MaindoorIntent` implementation. The `MaindoorIntent` structure conforms to the `SetValueIntent` protocol, which contains a value we can set. In this example, the value is from the `Bool` type, which we can use to perform the desired operation.

Adding a control widget to our app involves similar practices that we saw when we added a home screen widget and app intents that allow us to share code between widgets and other app components.

Summary

Widgets are enjoyable and fun UI elements we can work with in iOS development. They provide sleek UI, great animation, and a glanceable user experience. We have seen that each iOS version has added interesting new widget capabilities to make widgets more powerful than ever.

In this chapter, we learned about the idea of widgets, how to add widgets, creating a timeline, and adding user-configurable options. Also, we learned how to create custom animations and even add user interaction. *WidgetKit* has become a fascinating framework to work with. In the next chapter, we'll continue to discover how to improve the user experience, this time with SwiftUI animations.

6

SwiftUI Animations and SF Symbols

The previous chapter dealt with a delightful topic – widgets. Their aesthetic level is both enjoyable and effective, which makes working with them fun and easy. Now, we will take that feeling even further with SwiftUI animations.

Animation is a crucial topic in iOS development, as it enriches the experience and makes our app more intuitive and enjoyable to use. If you are used to UIkit animations, you will notice that SwiftUI animations take a different approach than UIkit, providing a declarative API to animate state changes.

With these new challenges also come opportunities that ensure our logic state and UI are always aligned.

In this chapter, we will do the following:

- Discuss the importance of animations
- Understand the SwiftUI animation concept
- Perform basic animations with the view modifier and the `withAnimation` function
- Perform advanced animations such as transitions and keyframe animations
- Animate SF symbols

Explaining why we need animations sounds weird and some may raise eyebrows about this topic. So, our first mission is to take this topic off the table before we move one pixel on the screen. So, let's answer the following question – why do we need to care about animations?

Technical requirements

For this chapter, it's essential to download Xcode version 16.0 or higher from the App Store.

Ensure that you're operating on the most recent version of macOS (Ventura or newer). Just search for Xcode in the App Store, choose the latest version, and proceed with the download. Open Xcode and complete any further setup instructions that appear. After Xcode is completely up and running, you can begin.

Download the sample code from the following GitHub link:

`https://github.com/PacktPublishing/Mastering-iOS-18-Development/tree/main/Chapter%206`

The importance of animations

Some of you may think that executing animations is mainly for fun and doesn't really impact an app's usability. But the truth is that animations play a crucial role in enhancing user engagement and interface design, especially in mobile applications. Here are a few benefits of using animations:

- First, animations provide **visual feedback** in response to users' actions – a button that grows when a user taps on it helps them know that they touched the right place
- Animations can also provide **guidance and navigation** – transitions between pages indicate whether we move "forward" with our flow or backward
- Animations also help in **error handling** – we can animate error messages and general issues and reduce a user's frustration
- Most importantly, in many cases, animations are part of the **app branding and uniqueness** and provide that special touch that strengthens the link between a user and an app

Now that we understand the importance of animation, let's see how SwiftUI's declarative approach aligns with that concept.

Understanding the concept of SwiftUI animations

For a developer coming from UIkit and taking their first steps in SwiftUI, the concept of writing animations in a declarative framework could feel a little awkward. After all, performing animations in UIkit was extremely simple – all we had to do was respond to some event and change some view properties within an animation closure. Here's a simple example of how to fade out a view in UIkit:

```
UIView.animate(withDuration: 2.0, animations: {
        sampleView.alpha = 0.0
    }) { (finished) in
}
```

In this example, we modify the alpha level of `sampleView` inside a `UIView` animation closure.

While this looks pretty simple, it comes with a significant drawback – the need to sync the animation action to the screen state. The `sampleView` component is now hidden – but does that mean that our view model or any other logic we incorporated in our screen is updated? This update is our responsibility. While this is a general *UIkit* problem, syncing between the view and the state can worsen when working with animations.

However, in *SwiftUI*, the screen state is always synced with the UI, and that's true for animations as well. The basic concept of SwiftUI animations revolves around the idea of animating changes to the view state, including properties such as position, size, opacity, and rotation.

There are several ways of implementing animations in SwiftUI; some are truly simple, while others let us deliver advanced and complex animations.

Let's warm up and start with some basic animations.

Performing basic animations

The fundamental way to understand how SwiftUI animations work is by associating a state value with a particular animation flow.

There are three ways of performing basic animations in SwiftUI:

- **Using the** `animation` **modifier** – adding an animation to a specific view
- **Using the** `withAnimation` **global function** – performing animation by changing several states
- **Using** `animation()` **method** – attaching an animation to a binding value

Developers usually get confused and think there's some duplication here – separate ways to perform the same functionality. But the truth is that all three serve different purposes and needs. It's up to us to decide the suitable way, according to our specific code structure and flow. Sometimes, you want to perform a particular animation to a specific view; occasionally, it is a shared experience with several views. Understanding the different use cases can help us decide how to perform an animation correctly.

Let's start by adding an animation to a specific view.

Using the animation view modifier

The animation view modifier goal is to add animation to a specific view when a certain value changes. Here's an example of using the animation view modifier:

```
struct UsingAnimationModifier: View {
    @State var width: CGFloat = 50
    @State var height: CGFloat = 50
    var body: some View {
        ZStack {
            Circle()
                .frame(width:width, height:height)
                .foregroundColor(.blue)
```

```
            .animation(.easeIn, value: width)
            .onTapGesture {
                width += 50
                height += 50
            }
        }
    }
}
```

The preceding code changes the circle size by adding 50 points to its width and height, and it does that by using the animation view modifier. Note that the animation view modifier has a value parameter – the value the animation modifier monitors for changes. In this case, we use the `width` state variable.

The animation view modifier is great for changing a specific view when a specific value changes. However, there are cases where this approach can be confusing. In this case, we define the animation in a specific place in the code but perform the change in another location. Moreover, using the animation view modifier can be cumbersome if we want to perform multiple animations.

If we want to perform multiple changes, we can use the `withAnimation:` function. Let's see how to utilize it.

Using the withAnimation function

In its basic form, the `withAnimation:` function takes a closure as a parameter and animates any changes made within that closure. Usually, that's done with a trigger to an event. Let's see a simple code example:

```
struct UsingWithAnimationFunction: View {
    @State var greenCircleYPosition: CGFloat = 400
    @State var redCircleYPosition: CGFloat = 800
    
    var body: some View {
        VStack {
            ZStack {
                Circle()
                    .size(width: 100.0, height: 100.0)
                    .foregroundColor(.green)
                    .position(x: 400, y:
                        greenCircleYPosition)
                
                Circle()
                    .size(width: 100.0, height: 100.0)
                    .foregroundColor(.red)
                    .position(x: 200, y:
```

```
                        redCircleYPosition)
                }

                Button("Animate") {
                    withAnimation {
                        greenCircleYPosition =
                            greenCircleYPosition == 400 ? 800 :
                            400
                        redCircleYPosition = redCircleYPosition
                            == 800 ? 400 : 800
                    }
                }
            }
        }
    }
}
```

This code example simultaneously animates the positions of two circles when a button is tapped. We can see that, unlike the animation view modifier, by using the `withAnimation:` function, we bind the change to the animation more clearly and simply.

Another advantage that `withAnimation:` has is the ability to execute a **completion code** once an animation ends.

Let's take a look at the following code example:

```
struct WithAnimationCompletionBlock: View {

    @State var yPos: CGFloat = 300
    @State var isReset: Bool = false

    var body: some View {
        VStack {
            Circle()
                .foregroundColor(.blue)
                .frame(width: 50, height:50)
                .position(x: 200, y:yPos)
            Button(isReset ? "Reset" : "Start") {
                withAnimation {
                    if isReset {
                        yPos = 300
                    } else {
                        yPos = 500
                    }
                } completion: {
```

```
                        isReset.toggle()
                    }
                }
            }
        }
    }
}
```

The code creates a blue circle and a button saying **Start**. Once the user taps the button, the circle animates its position, and at the end, the button title changes to **Reset**. Then, tapping the button brings back the circle, and at the end of the reverse animation, the button title returns to **Start**.

Completion blocks in animations are essential to sync flow stages. For example, collapsing a side drawer and navigating to a new screen at the end is an excellent example of completion block usage.

Now, it's time to bring some more life to our animation.

Bringing some life to our animations with spring animations

If you have tried out the code examples you have seen so far, you have probably noticed that the animations ran smoothly but were a little bit, well, boring. That's because the animations ran linearly and were not that interesting.

Try adding the following parameter to the previous example:

```
withAnimation(.bouncy(extraBounce: 0.3)) {
        if isReset {
            yPos = 300
        } else {
            yPos = 500
        }
    } completion: {
        isReset.toggle()
    }
}
```

In this example, we added .bouncy(extraBounce: 0.3) to our withAnimation function. Running the code shows the same animation as before, but now, the circle bounces when it reaches the end. It is a small but significant addition – the bounce effect adds a realistic touch to our animation and can improve user engagement.

There are several interesting visual transitions we can add to our animations. For example, we can make the bouncing smoother using the .smooth function:

```
withAnimation(.smooth(extraBounce: 0.3))
```

We can also make the bouncing snappier by making the animation faster with a small bounce amount:

```
withAnimation(.snappy)
```

It is recommended to look at Apple's documentation to discover more visual transitions that we can apply easily to our animations: https://developer.apple.com/documentation/swiftui/animation.

So far, we have performed very basic animations. But modern apps require modern experiences. Let's move on to some more ways to create advanced animations.

Performing advanced animations

We mentioned that transitions are great for guidance and navigation, and part of that concept is providing clarity about incoming and leaving views from our canvas. Sliding a view from the bottom can provide a sense of a drawer being opened and closed, and scaling a view can visually represent the progress of an ongoing process.

So far, we have discussed how to animate views from one state to another. Now, we will explore transitions – a way to animate views when they appear or disappear.

Performing transitions

Implementing a view transition is easy – we have some nice built-in transitions to choose from, and if that's not enough, we can also create a custom transition.

Let's start with some basic, built-in transitions.

Implementing built-in transitions

To add a transition, we should use the `transition` modifier with the specific view we want to animate, triggering it using the `withAnimation` function we learned about in the *Using the withAnimation function* section.

Here's a simple example of a slide in transition:

```
struct BuiltInTransitionsView: View {
    @State var showSlideText: Bool = false

    var body: some View {
        VStack {
            Button("Slide in text") {
                withAnimation {
                    showSlideText.toggle()
                }
```

```
            }
            if showSlideText {
                Text("Hello, slided
                    text").transition(.slide)
            }
        }
    }
```

The code example consists of `VStack` with a button and text. We also have a state determining whether the text is visible or hidden.

Tapping on the button reveals the text using the `withAnimation` function. But the text also has a transition view modifier that describes how it is supposed to appear – in this case, using a sliding-in transition.

The transition view modifier describes how the view appears and how it is supposed to disappear.

The `slide` transition inserts the view by moving it from the leading edge and removing it toward the trailing edge. Note that the slide transition directions cannot be changed, and they are set by the SwiftUI framework. However, there are several more transitions we can use to achieve our desired behavior:

- `move`: Moves the view in/from a specific edge:

    ```
    Text("Hello, moved text")
                    .transition(.move(edge: .bottom))
    ```

- `scale`: Scales the view in a specific amount and from a specific anchor:

    ```
    Text("Hello, scaled text")
                    .transition(.scale(scale: 0.5, anchor:
    .center))
    ```

- `opacity`: Performs a "fade in/out" effect on the view:

    ```
    Text("Hello, opacity text")
                    .transition(.opacity)
    ```

These types of transitions are well documented in the Apple website and SDK, and we can also try them using the chapter's GitHub repository.

It's important to note that we can use these transitions to show and hide animations. Yet, in some cases, we might prefer a different animation for hiding compared to showing. Having a different animation for hiding and showing is called an **asymmetric transition**. Let's see a code example for that:

```
Text("Text scaled in. Now it will slide out")
  .transition(.asymmetric(insertion: .scale, removal:
    .slide))
```

This code example performs a `scale` animation for the insertion of text and a `slide` animation for the removal of text.

Sometimes, we may want to combine several animations. For example, we may want to scale and slide at the same time. We can do that using the `combined` function:

```
Text("Scale and slide")
                .transition(.scale.combined(with:
                   .slide))
```

We can even combine a combined transition!

```
.transition(.scale.combined(with: .slide.combined(with:
    .opacity)))
```

However, if things become too complicated, it could be a sign that we should build a custom transition.

Creating a custom transition

Building **custom transitions** gives us complete control and flexibility of how transitions work and is useful when other compound transition methods don't provide the expected results.

The idea of building a custom transition is built around providing two view modifiers:

- One that represents the *identity* state of the view (before we started the transition)
- One that represents the *active* state of the view (after the transition)

Both view modifiers must be of the same type so that SwiftUI has the same properties to transition.

Let's create a custom transition that takes a view and inserts it with rotation, opacity, and scale.

We will start by creating a view modifier that handles all the three properties:

```
struct ViewRotationModifier: ViewModifier {
    let angle: Angle
    let opacity: CGFloat
    let scale: CGFloat

    func body(content: Content) -> some View {
        content
            .rotationEffect(angle)
            .scaleEffect(scale)
            .opacity(opacity)
    }
}
```

The `ViewRotationModifier` view modifier receives three properties, `angle`, `opacity`, and `scale`, and applies them to the content. This view modifier is like any view modifier we're accustomed to.

Now, we can build our custom transition. If we look at the built-in transitions we covered in the previous *Implementing built-in transitions* section and their code's documentation, we can see that they are from the type `AnyTransition`. `AnyTransition` is a struct that describes a SwiftUI transition between two states.

Let's build our `rotate` `AnyTransition`:

```
let rotate = AnyTransition.modifier(
    active: ViewRotationModifier(angle: .degrees(360),
      opacity: 0.0, scale: 0.0),
    identity: ViewRotationModifier(angle: .degrees(0),
      opacity: 1.0, scale: 1.0)
)
```

The `AnyTransition` struct we created receives the `active` and `identity` view modifiers, each with different parameters.

We can use the new transition in the same way as the built-in transitions:

```
struct CustomizedTransitionView: View {
    @State private var showRectangle: Bool = false
    
    var body: some View {
        VStack {
            Spacer()
            
            if showRectangle {
                Rectangle()
                    .frame(width: 100, height: 100)
                    .foregroundColor(.blue)
                    .transition(rotate)
            }
            
            Spacer()
            
            Button("Insert Rectangle") {
                withAnimation {
                    showRectangle.toggle()
                }
            }
        }
    }
}
```

The preceding code creates a rectangle and a button. Tapping on the button toggles the `showRectangle` state variable, which reveals the rectangle using our new transition.

So far, we have discussed great animations that were pretty simple and short. However, if we want to provide more sophisticated animations that may require multiple stages and different timing, `AnyTransition` structure is insufficient. For much more advanced animations, we should try to implement keyframe animations.

Executing keyframe animations

The idea of **keyframe animations** in SwiftUI is similar to how they are implemented in UIkit.

With keyframe animations, we declare different changes in different properties over time. There are four primary components in keyframe animations:

- **Animations properties**: A structure that defines the changes we want to perform during the animation phase. For example, the `AnimationsProperties` struct can define the opacity, scale, or color in different animation phases.
- `KeyFrameAnimator`: The keyframe animator defines the different animation tracks we have and what happens with the view in each track.
- `KeyframeTrack`: Each track handles a different animation property and defines the various phases (key frames) for that property. Tracks work in parallel with each other. A keyframe animator can have multiple tracks.
- `KeyFrame`: Defines a single change for a specific property within the keyframe track.

With these four primary components, we can build amazing and complex animations. Let's build our first keyframe animation with SwiftUI, but we'll start by explaining the concept behind keyframe animations.

Understanding a keyframe animation

Describing a keyframe animation can be slightly confusing at first, mainly because it is a way to create complex animations. Let's try to explain it in a diagram (*Figure 6.1*):

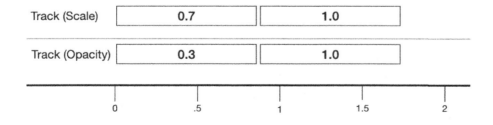

Figure 6.1: A key frame animation as a diagram

Figure 6.1 shows two tracks – scale and opacity – positioned on a timeline. In each track, we see two keyframes. The number inside each keyframe describes the value, and the keyframe length describes its duration. For example, in the scale track, we have two keyframes – the first sets the scale to 0.7, and the second brings it back to 1.0. We can also see that the durations of both the scale and opacity tracks are equal.

If you think that that resembles a video editing application such as *iMovie* or *Premiere*, that's because it is based on the same concept.

Let's try to create a breathing animation using the concept of keyframe animation. A breathing animation mimics the way something breathes, such as a balloon slowly inflating and deflating.

Let's see how to do that in code:

```
struct AnimationProperties {
    var scale = 1.0
    var opacity = 1.0
}

struct KeyFrameAnimations: View {
    var body: some View {
        Circle()
            .foregroundColor(.red)
            .frame(width:100, height:100)
            .keyframeAnimator(initialValue:
                AnimationProperties(), repeating: true) {
                content, value in
                content
                    .opacity(value.opacity)
                    .scaleEffect(value.scale)

            } keyframes: { _ in
                KeyframeTrack(\.scale) {
                    CubicKeyframe(0.7, duration: 0.8)
                    CubicKeyframe(1.0,
                            duration: 0.8)
                }

                KeyframeTrack(\.opacity) {
                    CubicKeyframe(0.3, duration: 0.8)
                    CubicKeyframe(1.0, duration: 0.8)
                }
            }
    }
}
```

The code example seems long! However, upon closer examination, we can see that it is not that complex and contains the different components we discussed earlier.

Let's explain what we've done here:

1. We created a circle and added a view modifier called `keyframeAnimator`, which handles the general animations. We initialized it with the `AnimationProperties` struct that holds the properties we want to modify during the animation phases, and we defined that animator to repeat by passing `true` in the corresponding parameter.
2. The animator has another closure parameter with the content view and the value. That's where we can *modify our view* according to the animation properties. In this example, we changed the view opacity and scale.
3. Right after the closure, we define our tracks. We have two properties we want to change over time, so we've created two tracks – one for scale and one for opacity. Because we wanted a *breathing* animation, we've created two keyframes – one for exhaling (scale down and reduce opacity) and one for inhaling (scale up and increase opacity).
4. We can see that each one of the frames is declared as `CubicKeyframe`. Before we explain what `CubicKeyframe` means, let's talk about keyframes, which are fundamental concepts in animations.

A keyframe specifies an object's state at a particular point in time. The animator's responsibility is to perform the animations between these keyframes. In a way, it's like animating a state change, but in this case, we define the different modifications upfront.

In the case of SwiftUI's `keyframeAnimator`, the keyframes align with the concept of states – each keyframe defines a change in a specific property over time.

In SwiftUI, we have different types of keyframes, each representing a different experience:

- `CubicKeyframe`: This is the keyframe we used in our code example. `CubicKeyframe` provides a smooth transition to the next keyframe while computing something called **Catmull-Rom splines**. Catmull-Rom splines are curves used in computer animations to provide smooth movement.
- `SpringKeyframe`: This represents a transition that emulates a spring experience, including a bouncy effect.
- `MoveKeyframe`: This type of keyframe modifies the given value immediately.
- `LinearKeyframe`: This keyframe animates the change without a defined curve and, instead, does that in a simple linear interpolation.

SwiftUI is intelligent enough to smoothly handle the combination of different keyframes on the same track. For example, let's see what happens when we define velocity on one of our keyframes:

```
CubicKeyframe(0.5, duration: 0.2, startVelocity: 0.5,
    endVelocity: 0.8)
CubicKeyframe(0.7, duration: 0.5)
```

We can see that the end velocity of the first keyframe is 0.8. However, we haven't defined any initial velocity for the second keyframe. In this case, the second keyframe's `startVelocity` value will be the end value of the previous keyframe, which means 0.8.

Now, let's discuss another crucial aspect of keyframe animations – animation duration.

Handling keyframe animation duration

The keyframe animator is a hierarchal structure with three levels – the animator, the tracks, and the keyframe. This means that different keyframes can have different durations, and these duration values don't always add up nicely. That makes duration management complex, especially for long and intricate animations.

How do we ensure that all the keyframe durations are always aligned with each other and maintain the same scale? The answer is to use relative duration, not absolute duration.

An absolute duration specifies the exact time an animation should take, regardless of the initial state, or without comparing it to the other keyframes.

Conversely, relative duration reflects the duration time, considering the total animation duration. For example, if the relative duration is 0.5 and the total animation duration is 3 seconds, the actual keyframe duration would be 1.5 (0.5 * 3.0 seconds).

By using relative duration, we can establish an animation's overall duration and allocate specific durations for each keyframe, relative to the total duration.

Let's take our "breathing" example and try to implement relative duration:

```
let duration: TimeInterval = 1.8

var body: some View {
    Circle()
        .foregroundColor(.red)
        .frame(width:100, height:100)
        .keyframeAnimator(initialValue:
          AnimationProperties(), repeating: true) {
            content, value in
              content
                .opacity(value.opacity)
```

```
                    .scaleEffect(value.scale)

    } keyframes: { _ in
        KeyframeTrack(\.scale) {
            CubicKeyframe(0.7, duration: 0.5 *
                duration)
            CubicKeyframe(1.0,
                duration: 0.5 * duration)
        }

        KeyframeTrack(\.opacity) {
            CubicKeyframe(0.3, duration: 0.5 *
                duration)
            CubicKeyframe(1.0, duration: 0.5 *
                duration)
        }
    }
}
```

In this code example, we have a keyframe animation with two keyframes, similar to our previous example. The first keyframe handles the scale animation, and the second handles the opacity.

We can see that we have a total duration variable, currently set to `1.8`. With each keyframe, we set the duration relative to that value. In this case, it is `0.5` of the total duration, but this can vary from one example to another.

Relative duration can help us set a dynamic overall duration time and change it according to our needs, even at runtime.

SwiftUI animations are extremely powerful and easy to use, and keyframe animations make them even more powerful by allowing us to build complex animations with multiple steps and durations.

However, in many cases, animating views is one of the many challenges that app developers face. After all, animating simple shapes such as a rectangle or a circle isn't always what we desire. So, what about the assets? Fortunately, the iOS SDK contains a fantastic resource called SF Symbols. Let's explore it now.

Animating SF Symbols

SF Symbols is a library that contains over 5,000 symbols that developers can integrate within their text, using the *San Francisco* font.

Don't be confused – SF Symbols are not emojis. Emojis are meant to express feelings and emotions within text. Conversely, SF Symbols are excellent replacements for icons that represent states, actions, and tools.

Here's a basic example of displaying a clock alarm symbol with text next to it:

```
var body: some View {
    HStack {
        Image(systemName:
            "alarm.waves.left.and.right.fill")
        Text("Alarm")
    }.font(.system(size: 30))
}
```

We can see no surprises here – we use a basic `Image` view with the `systemName` parameter to provide the image name.

As mentioned earlier in this section, there are thousands of symbols available. To get the full symbols catalog, we need to download a Mac application called *SF Symbols* (what a coincidence, uh?) from `https://developer.apple.com/sf-symbols/`.

The app is simple to use, as we can see in *Figure 6.2*:

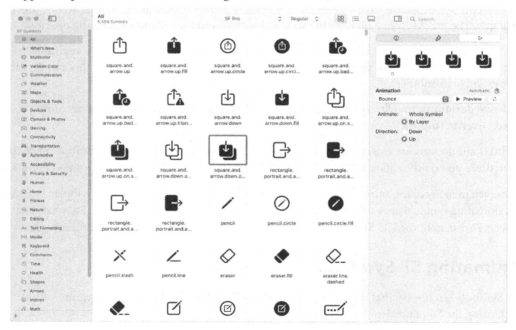

Figure 6.2: The SF Symbol Mac app

By exploring the SF Symbol app, we can see how the symbols differ from emojis. They are not only vector illustrations (meaning they can scale to any size) but also built as layers.

To understand why the SF symbols contain layers, try to perform a bounce animation using the app. Doing so lets us see how the layers create a *sense of depth*, making them bounce at different intervals.

Other than the bounce effect, SF Symbols supports other effects such as pulse, scale, and replace. We can perform the same animations in our SwiftUI code using the `symbolEffect` view modifier:

```
struct SFSymbolsAnimationView: View {
    @State private var animate = false
    var body: some View {
        HStack {
            Image(systemName:
                "alarm.waves.left.and.right.fill")
                .symbolEffect(.bounce, options: .repeating,
                    value: animate)
            Text("10:30")
        }.font(.system(size: 40))
            .onTapGesture {
            animate = true
        }
    }
}
```

The `symbolEffect` view modifier has several parameters. The first is the `effect` type, the same as those found in the SF Symbol app. The second parameter is `options` – we can make the effect repeat itself or even set its speed.

The third parameter is the `value` parameter – the state variable that triggers the animation. In this case, we trigger the animation by tapping on the `HStack` view that contains both the symbol and the attached text.

To read more about SF Symbols, it is recommended to visit Apple's website: `https://developer.apple.com/sf-symbols/`.

Even though this chapter mainly concerns SwiftUI animations, there is much more to SF Symbols than just animations, such as supporting multiple colors. Let's see how we can modify different symbol colors.

Modifying symbol colors

The fact that SF Symbols are built with different layers helps not only with animation but also with coloring them.

Let's take, for instance, the *two persons waving* symbol:

```
Image(systemName: "person.2.wave.2")
```

Figure 6.3 shows what the symbol looks like:

Figure 6.3: The person.2.wave.2 symbol

We can see two different types of image components – on the one hand, two people, and on the other hand, their waves. So, unlike a regular image, we can set one color for the people and another for the waves.

Every SF Symbol has a **primary** and **secondary** color, and SwiftUI knows how to color it accordingly.

For example, let's set a primary color of brown and a secondary color of blue. We will use the `foregroundStyle` view modifier for that:

```
Image(systemName: "person.2.wave.2")
            .foregroundStyle(.brown, .blue)
```

There are symbols that even have a third color, such as in the case of the three-person symbol (*Figure 6.4*):

Figure 6.4: The person.3.sequence.fill symbol

To use the third color, we just need to add one more color as a parameter:

```
Image(systemName: "person.3.sequence.fill")
            .foregroundStyle(.red, .blue, .brown)
```

Anyone who has had to manage multi-color icons knows the complexity of supporting different themes and colors, especially when we need to animate them.

So, we know how to add an SF Symbol, animate it nicely, and color it. However, we can also use vector multi-layer symbols, which is known as localization.

Localizing our symbols

Localizing our apps is a crucial topic today, more than ever. However, how many of us pay attention to icon localization and try to adjust them according to the app layout direction?

The excellent news about SF Symbols is that they can adjust to the current app locale. The even better news is that we can force them to do that if we want.

But why do SF Symbols even need to support localization?

Let's take the `arrowshape.turn.up.forward` SF Symbol (*Figure 6.5*):

Figure 6.5: The arrowshap.turn.up.forward SF Symbol

The forward icon arrow points to the right, which fits nicely in **LTR (Left-to-Right)** layout views. But what about **RTL (Right-to-Left)** layouts, such as in Hebrew or Arabic localized applications?

Well, in this case, we will have to flip the icon direction. With SF Symbol, this adjustment is done automatically for us.

Moreover, we can set the icon localization regardless of the view settings, using the `environment` view modifier:

```
Image(systemName: "arrowshape.turn.up.forward")
    .environment(\.layoutDirection, .rightToLeft)
```

In the preceding code, we force the SF Symbol to have an RTL layout direction, which flips the forward arrow to the left direction.

Having localization support doesn't stop with layout direction. Some symbols even change their look according to the current locale.

For example, let's take the `character.book.closed` SF Symbol (*Figure 6.6*):

Figure 6.6: The character.book.closed SF Symbol

In the case of the symbol in *Figure 6.6*, we can see that in addition to its layout direction (LTR), it also has a letter on it.

In the case of the Hebrew locale, not only does the symbol's direction change but also the letter (*Figure 6.7*):

Figure 6.7: The character.book.closed SF Symbol in a Hebrew locale

We can force the symbol to retrieve a specific locale using the `environment` view modifier, similar to the layout direction:

```
Image(systemName: "character.book.closed")
            .environment(\.locale, .init(identifier: "he"))
```

To sum up, SF Symbols contain so much power and valuable features. Trying to support standard icons in different environments, such as locales and themes, can be a hassle, and animating them without creating a dedicated image sequence is almost impossible. So, getting all these features for free is like a massive present from Apple engineers.

Summary

iOS animations are like salt – they can enhance the user experience, but too much is overwhelming.

The great thing about SwiftUI animations is that they are aligned to the screen state because of the declarative implementation. However, it's a significant change to how they work in UIkit.

Because of that, in this chapter, we went from understanding the basic concepts and performing fundamental animations to custom transitions and keyframe animations, and we even discussed a great present that Apple gave us, SF Symbols.

Now, we should be able to easily animate changes on our screen in a meaningful and expressive way!

In our next chapter, we'll explore enhancing user engagement using a built-in solution – TipKit.

7
Improving Feature Exploration with TipKit

In the previous chapter, we learned about SwiftUI animations. We know now that SwiftUI animations are a great way to teach users how to use our app.

However, sometimes, it's not enough, and we need more than fancy animations. This is where TipKit comes in. TipKit's goal is to provide a solution for another important topic: feature exploration. Feature exploration affects our app users' engagement and usage, eventually affecting user satisfaction and experience.

In this chapter, we will cover the following topics:

- Learning the importance of tips in a mobile app
- Adding a new tip – inline and popover
- Customizing our tip's feel and look
- Supporting tip actions
- Defining display rules for our tips
- Grouping tips using TipGroup
- Adjusting display frequency

Now, let's start with the fundamental question – why do we need TipKit?

Technical requirements

For this chapter, it's essential to download Xcode version 16.0 or higher from the App Store.

Ensure that you're operating on the most recent version of macOS (Ventura or newer). Just search for Xcode in the App Store, choose the latest version, and proceed with the download. Open Xcode and complete any further setup instructions that appear. After Xcode is completely up and running, you can begin.

Download the sample code from the following GitHub link: `https://github.com/PacktPublishing/Mastering-iOS-18-Development/tree/main/Chapter%207`

Learning the importance of tips

One of the challenges of creating an app for a small screen, such as a smartphone screen, is to provide ways for the user to explore valuable features. Making users use more features is part of improving user engagement – measuring how much the user is actively involved and connected to our product.

That feature exploration is a real challenge. On the one hand, we aim to create a clean and straightforward user interface and, on the other hand, we aim to add more features that can be extremely useful to our users.

Every product manager struggles with this challenge – sometimes, the solution is to create a **What's new** popup, send a marketing email, or add more information to the in-app FAQ screen.

One of the most valuable techniques is to provide a tip – a small text box that pops up in the right place at the right time to explain a new feature and can even add an action to help the user use it.

Let's try to drill down a bit and discuss the basics of tips in Apple's TipKit framework.

Understanding the basics of TipKit

Some may think that the primary challenge with displaying tips is creating views that contain relevant information and presenting it.

However, if that were the case, we wouldn't need a whole framework. Instead, we should consider TipKit a complete system.

Let's look at *Figure 7.1*:

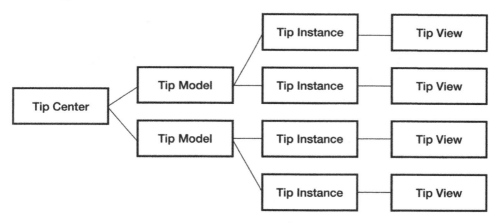

Figure 7.1: Tip infrastructure

Figure 7.1 presents the essential components of the structure of TipKit in iOS. First, there is the **tip center**, a singleton that manages all the tips' appearances in the app. The tip center has several responsibilities:

- Ensures that a tip stops appearing once invalidated or dismissed by the user
- Triggers the tips such that tips don't overlay each other
- Displays tips according to specific rules

Right after the tip center, we have the **tip model** – a structure representing a specific tip declaration. Based on the tip model, we can create and display an instance using a **tip view** – a visual representation of the tip.

The TipKit infrastructure looks more complex than it is – many iOS frameworks work with some framework center, models, and views. However, the idea here is to show you that while TipKit provides visual components, its core functionality lies in the rules that determine when and where they appear.

Now, enough of the theoretical introduction. Let's see what tips look like!

What do tips look like?

The result of adding a tip is a view that shows the new feature to the user (*Figure 7.2*):

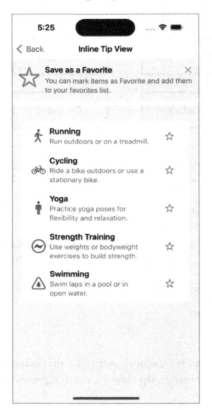 

Figure 7.2: Two different ways to display tips in TipKit

Based on *Figure 7.2*, we can see that there are two ways of displaying a tip:

- **Inline tip**: Inline tips embed themselves as part of the screen layout, and their appearance modifies and pushes the other views accordingly. Inline tips are a great fit in **VStacks** or **Lists**, and we can view them without interfering with the screen interaction.
- **Popover tip**: Unlike inline tips, popover tips appear above the current screen, mostly linked to a button or another control. With popover tips, the user must dismiss the tip or perform its action to continue with the app. Also, we cannot display multiple popover tips at the same time.

At first glance, the two ways to display tips are just a matter of design. Showing a popover tip is not only a different experience but also a different use case. In the inline tip, we include the view in a non-intrusive way that allows the user to discover new features gradually. In contrast, the popover tip is excellent for contextual help or complex features requiring guidance. Remember that adding more views to our device's small screen can sometimes be overwhelming, and we must make this decision carefully.

Let's add our first tip and convince users to use our new **mark as favorite** feature.

Adding our first tip

Even though we are excited and can't wait to add our first tip, we still need to do a small amount of preparation first, and that's setting up our persistent store for the tips state.

Why do we need to set up a persistent store? TipKit needs to manage the tips' state over time, even after we close our app. This is because we don't want our tips to show up again if the user (or the app) decides to dismiss them.

We can set up the persistent store using a static function called `configure()`:

```
import TipKit

@main
struct MyApp: App {

    init() {
        try? Tips.configure()
    }
}
```

In our code example, we can see that we call the `configure` function part of the app initialization process because we need TipKit to have all its information once the first screen is loaded.

We can also share the tips' states store with more of our apps and extensions by defining it in a group container:

```
try? Tips.configure([
    .datastoreLocation(.groupContainer(identifier:
      "MyAppGroupContainer"))])
```

In this example, we configured the tips' state data store in a group container called `MyAppGroupContainer`. Don't worry – from the app's perspective, the user experience will stay the same.

> **What is a group container?**
> A group container is a directory shared among multiple apps and extensions within the same app group. It allows us to share data between our apps.

Our next step is to define our `Tip` model.

Defining our Tip model

The `Tip` model (based on the **Tip protocol**) defines our tip's behavior and appearance.

Let's see a simple tip declaration:

```
struct MarkAsFavoriteTip: Tip {
    
    var id: String { "InlineTipView"}
    
    var title: Text {
        Text("Save as a Favorite")
    }
    
    var message: Text? {
        Text("You can mark items as Favorite and add them
            to your favorites list.")
    }
    
    var image: Image? {
        Image(systemName: "star")
    }
}
```

In this code example, we declared a struct named `MarkAsFavoriteTip`, which conforms to the `Tip` protocol. We can see that `MarkAsFavoriteTip` has several properties. The title, message, and image define the content of the tip view, as we can see in *Figure 7.3*:

Figure 7.3: The Save as a Favorite tip view

In *Figure 7.3*, we can see the tip view with all the content we declared in `MarkAsFavoriteTip`. Now, let's see how we add this tip to our SwiftUI view:

```
struct InlineTipView: View {

    var tip = MarkAsFavoriteTip()

    var body: some View {
        VStack {
            TipView(tip)
            List(workouts) { workout in
                WorkoutView(workout: workout)
            }
        }
    }
}
```

This code example has a SwiftUI view that presents a list of workouts. To display a tip on top of the list, we created an instance of the `MarkAsFavoriteTip` structure from earlier and then a `TipView` view, passing that tip instance.

Figure 7.4 shows what it looks like:

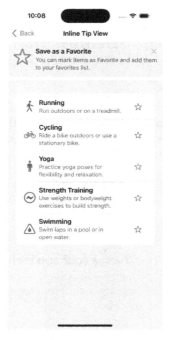

Figure 7.4: A workout list with an inline tip view

Figure 7.4 shows how the tip fits nicely on the screen, pushing the list down the `VStack` view. Tapping on the tip's close button removes the tip from the screen and pushes the list up to take its place.

Pretty simple, huh? Now, let's see how to add a popover tip.

Adding a popover tip

As mentioned, a popover tip serves different use cases than an inline tip. It blocks the user's interaction with other elements on the screen and is more contextual. In a popover tip, we link the popover view to a specific control on the screen – usually a button or a toggle.

The way we add a popover tip is by using a view modifier called `popoverTip`, passing our tip instance (`tip`) (just like in an inline tip) and an optional arrow direction:

```
struct PopoverTipView: View {
    var tip = PopoverTip()
    var body: some View {
        List {
            // some list information
        }
        .navigationTitle("Popover Tip")
        .toolbar(content: {
            Button("Settings", systemImage: "gearshape") {
            }
            .buttonStyle(.plain)
            .popoverTip(tip, arrowEdge: .top)
        })
    }
}
```

Our code example shows a similar pattern to the one we saw in the inline tip. We create a tip instance, and this time, we add the tip to our screen by using a view modifier attached to the screen toolbar button. *Figure 7.5* shows how it looks:

Figure 7.5: A popover tip

What's nice about the popover tip is that we don't need to care about things such as positioning, depth, or creating the popover pointer – this is all done for us, similar to the popover view modifier.

We see that both inline and popover tips have a close button. Let's discuss it further because this is where we start to reveal the tip's real added value.

Dismissing tips

You are probably wondering how dismissing a tip view is related to the tip's added value.

We discussed that the real power of tips is not the UI layer but its presentation logic. Whenever we dismiss a tip, TipKit marks it as invalidated and won't show it again. TipKit also stores the invalidation status permanently, meaning it won't be displayed after the app relaunch.

Another way to dismiss a tip, besides closing it, is to invalidate it in code. Let's look at the inline tip from the code example earlier again (under *Defining our Tip Model*). The tip in that example helps the user explore the app's favorites feature. It also means that whenever the user marks one of the workouts as a favorite, we can assume that the tip is no longer needed and invalidate it ourselves.

To invalidate a tip, we need to call the tip's `invalidate()` function:

```
List(workouts) { workout in
            WorkoutView(workout: workout,
              onFavoriteButtonTap: {
                tip.invalidate(reason:
                  .actionPerformed)
              })
          }
```

In this code example, we call the `invalidate()` function whenever the user taps the favorite button.

Remember that SwiftUI is a declarative framework – the tip status is part of the view state, and SwiftUI refreshes the screen after a change.

Another thing we can see in that code example is the reason for invalidation. In this case, we send `actionPerformed` because this is precisely what happened – the user performed the action suggested by the tip.

Another question may arise at this point: how does TipKit know whether that specific tip has already been shown? Also, is there a way to reset the persistent data and show the tip again?

That's where the tip ID comes in.

Defining the tip ID

If you went over the code example under the *Defining our tip model* section, you may have noticed the following line:

```
var id: String { "InlineTipView" }
```

The `id` variable is part of the `Tip` protocol; we use that property to define a specific identifier for each tip. TipKit uses that identifier to manage the different tip statuses.

You can do a small experiment: create a small app with a tip (or use the code examples in our GitHub repository) and invalidate the tip. Relaunch the app and see that the tip doesn't show again. Now, modify the tip identifier to have a different name. Relaunch the app, and you'll see that the tip is visible again. Also, reinstalling the app (after deletion) will reset the local persistent store.

Another way to reset the local persistent store is to call the static `resetDatastore` function on app launch:

```
struct MyApp: App {
    init() {
        try? Tips.resetDatastore()
        try? Tips.configure()
    }
}
```

Notice that we call the `resetDatastore` function before the `configure` function.

The tip identifier is part of the `Tip` protocol, and in this example, the identifier is shared across all the struct instances:

```
var id: String { "InlineTipView" }
```

Since the identifier is shared, all the tip views based on the `struct` instances will be dismissed once you invalidate one of them.

In most cases, that's considered normal behavior and best practice. If the user marks a specific row as a favorite, they already know about this feature, even if it appears on another screen.

However, that's not always the case, so plan identifiers accordingly.

Now we know how to present a tip, both inline and popover. We also know how to dismiss it and even reset the persistent state. However, TipKit provides even more than that. Let's see how we can customize our tips.

Customizing our tips

So, TipKit provides an excellent infrastructure for presenting persistence-based tips in our app. However, the TipKit framework's developers knew that handling tips requires more thought than just invalidating an ordinary view with an image and two texts.

Let's see how we can customize tips to our own needs. We'll start with their appearance.

Customizing our tips' appearance

Unlike many other UI-based frameworks Apple has provided, TipKit lets us customize the tip views nicely. This may be because SwiftUI is a declarative framework, and expressing visual content becomes more natural. However, in the case of TipKit, Apple understood developers' need to align the TipKit design with their apps.

There are two ways to customize our tip's appearance. The first is to modify the tip's properties and apply basic changes without changing the tip's layout and components.

The second way is to implement a new tip view style, which allows you to fully control the tip's feel and look. Let's start with the first way: modifying the tip properties.

Modifying the tip properties

As I mentioned earlier, one of the great things about SwiftUI is its expressive framework, and we can use view modifiers to adjust the tip's appearance to our style.

Let's look at the tip's `title` property again:

```
var title: Text { Text("Save as a Favorite") }
```

Notice that we are not returning a `String` but a `Text` value, which is a SwiftUI view. This means we can modify its look to look like any other SwiftUI view.

For example, we can change the title text color by applying the `foregroundStyle` view modifier:

```
var title: Text {
    Text("Save as a Favorite")
        .foregroundStyle(.red)
}
```

The code example is straightforward: we took the text view and changed its look. Moreover, because we can build a `Text` view by compounding multiple text views, we can mix styles and colors:

```
var title: Text {
    Text("Save as a ")
        .fontWeight(.light) +
    Text("Favorite")
```

```
            .fontWeight(.bold)
            .foregroundStyle(.red)
}
```

In this example, we took our `Save as a Favorite` text and changed the `Favorite` text to be red and bold to distinguish it from the rest of the title.

We can also perform changes to the `Image` property, such as changing its colors or rendering mode:

```
var image: Image? {
      Image(systemName:
          "externaldrive.fill.badge.icloud")
          .symbolRenderingMode(.multicolor)
}
```

In *Chapter 6*, we learned that SF symbols have multiple layers so that we can apply different colors to different layers. In this example, we changed the rendering mode for our symbol to `multicolor`.

Modifying the tip properties is a great way to add a basic touch to our tip view user interface. However, we know how crucial design is in iOS apps, and sometimes, changing colors and font style just isn't enough. For this reason, we can use `TipViewStyle` for further customization.

Using TipViewStyle

The given tip view design only works if we need a different UI layout or an even more complex tip view. So, we must consider a different design pattern to meet that requirement.

One of the most critical development principles I love to mention is the separation of concerns principle – the idea that different components should have other responsibilities.

Some responsibilities are mixed up when we look at how the `Tip` protocol works. On the one hand, the Tip protocol structure defines the content of our tip – the title, message, and image. On the other hand, the structure also defines its design, which may be a different responsibility.

The fact that these responsibilities are mixed also limits how we design our tips – we can't define a new layout as part of the structure.

However, content and design are part of SwiftUI's nature and one of its strengths as a declarative language. Fortunately, we have a solution for that: **View Styles**. A View Style is a way to define the appearance of a view component.

Here's an example of defining a bordered button:

```
Button("Sign In", action: signIn)
  .buttonStyle(.bordered)
```

In this code example, we stay with the same view (`Button`) but apply a specific style.

In TipKit, we can also define our tip appearance by applying a custom view style:

```swift
struct ImageAtTheCornerViewStyle: TipViewStyle {
    func makeBody(configuration: Configuration) -> some
      View {
        VStack {
            if let title = configuration.title, let message
              = configuration.message {
                title
                    .multilineTextAlignment(.center)
                    .font(.title2)
                Divider()
                message
                    .multilineTextAlignment(.leading)
                    .font(.body)
            }
            HStack {
                Spacer()
                Image(systemName: "star")
            }
            .padding()
        }
    }
}
```

The View Style we just created takes a `Tip` view and returns a new view with the same content but a different layout and design. It even adds a new view component, such as a `Divider` and `Spacer` component. The magic happens in the `makeBody` function, which receives a `Configuration` parameter that contains all the tip information.

To apply our new View Style on a tip, we can use the `tipViewStyle` method:

```swift
TipView(tip)
    .tipViewStyle(ImageAtTheCornerViewStyle())
```

Now, our `TipView` view has our new custom style and layout, and it looks like this (*Figure 7.6*):

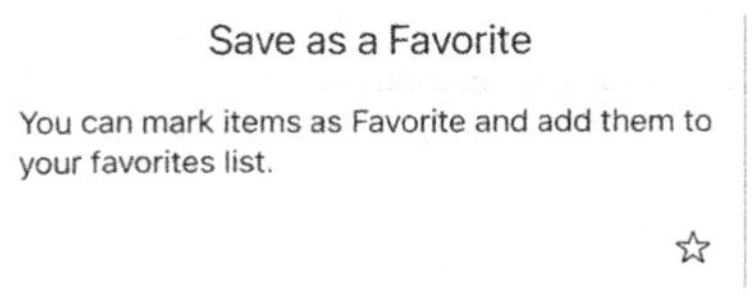

Save as a Favorite

You can mark items as Favorite and add them to your favorites list.

☆

Figure 7.6: Customizing our tip with TipViewStyle

It is a good practice to name the `TipViewStyle` protocol with a generic and descriptive name such as `ImageAtTheCornerViewStyle`, so it will be easier to share it with the rest of our tips.

Up to now, we've learned how to define a tip, present it in different places, and design it however we want. However, our journey with enriching our tips doesn't end here, as we can also add some user interactions by adding **actions**.

Adding actions

Actions are very valuable additions to tips. In many cases, our tips suggest that the user take an action – go to the settings screen, add a new task, or enter our app's new edit mode, for example. What's better than adding a button that performs that specific action inside the tip view?

Besides the title, message, and image, the tip protocol also contains an actions property – an array of structures describing the buttons our tip will display.

Let's see that property in an example:

```
struct ChangeEmailTip: Tip {
    var actions: [Action] {
        
        Action(id: "go-to-settings", title: "Go to 
          settings")
        
        Action(id: "change-now", title: "Change email now")
    }
}
```

The code example shows a `ChangeEmailTip` structure with two actions. (Notice that this tip is partial; assume that we already implemented the rest of the properties, such as `title` and `message`.)

Each action initialization function has two parameters: `title` and `id`. The `title` parameter represents the title that appears on the button. The `id` parameter describes the goal of this action, and we use it to determine which button the user tapped.

Figure 7.7 shows how the actions look in a popover tip:

Figure 7.7: Two actions in a popover tip view

Like the rest of the properties, TipKit decides what layout to present the actions in and how the buttons look.

Now that we have defined and presented the actions, let's see how we respond to user selection. Responding to an action is easy now that we have an ID for each action. The `popoverTip` view modifier we discussed in the *Adding a popover tip* section has an additional closure that handles action selection. Let's see a code example for that:

```
Button("Settings", systemImage: "gearshape") {
        gotoSettings = true
    }
    .buttonStyle(.plain)
    .popoverTip(tip, arrowEdge: .top) { action in
        if action.id == "go-to-settings" {
            gotoSettings = true
        }
    }
```

This code example shows the exact popover tip implementation, now with the closure that handles the selected action. Within the closure, we check for the action ID and perform the desired action (for example, navigate to the settings screen).

It's much nicer to add these ids in static constants for clarity:

```
struct ChangeEmailTip: Tip {
    // rest of the tip
 var actions: [Action] {
        
        Action(id: ChangeEmailTip.goToSettingsAction,
            title: "Go to settings")
        
        Action(id: ChangeEmailTip.changeEmailAction, title:
            "Change email now")
    }
    static let goToSettingsAction = "go-to-settings"
    static let changeEmailAction = "change-now"
}
...

.popoverTip(tip, arrowEdge: .top) { action in
        if action.id ==
            ChangeEmailTip.goToSettingsAction {
            gotoSettings = true
        }
    }
```

This code example shows how beautiful Swift can be when applying best practices!

Speaking of beautiful, we discussed how to design our tips using `TipViewStyle`, so we can also design our actions using the same technique:

```
List(configuration.actions) { action in
            Button(action:{
                // perform action
            }) {
                action.label()
            }
        }
```

In this code example, we create a list of buttons, each handling a different action. We need to add that code snippet to the `makeBody` function we learned about in the *Using TipViewStyle* section.

At this point, we've learned so much about tips! The good news is that we have more surprises up our sleeves. Let's reveal them and discuss the **rules** feature.

Adding tips rules

Throughout this chapter, we have focused mainly on the UI side of displaying tips so far. However, we already know that tips are more than just nice views – they must correspond to some app logic or states. For example, maybe there are tips that we present when the user is logged in. In a photo app, we can show a tip suggesting adding an album after the user takes a certain number of photos.

Tips must often be *made aware* of users' flows and states. That's why TipKit also contains a feature called rules.

There are two rule types:

- **Based on state**: Show or hide the tip based on a specific state. The user logged in, performed a particular action, and more.
- **Events tracking**: Show or hide the tip based on the number of events the user performed. For example, if the user entered a specific screen in the settings a few times in the past week, we could offer for them to create a shortcut for that screen.

Let's start with adding a rule based on a state.

Adding a rule based on a state

Creating rules based on a state is the common way to establish tip display logic. What is a state? A state can be an authentication state (is the user logged in?), unlocking goods, features usage, and more.

There are three steps for implementing a rule that is based on a state:

1. **Adding a parameter**: We need to add a variable on which the rule will be based.
2. **Define the rule**: The rules are defined inside the tip and should consider the parameter we discussed.
3. **Connect the parameter to the app logic**: If we want the rule to be based on our app's real state, we need to maintain and synchronize it with the app state.

Believe it or not, implementing a rule-based tip is even easier than it looks! Let's try to build a tip encouraging our users to use a premium-only feature such as changing the app theme.

Adding a parameter

The rule needs to rely on persistent data that the app can easily modify to track an app state. To do that, we use the `@parameter` macro to add a tracking state variable to our tip.

> **What is a macro?**
> A macro is a Swift feature that helps the compiler generate code based on the current code and parameters. You can read about macros in *Chapter 10*.

Let's add a parameter called `isPremiumUser` to track premium eligibility:

```
struct ChangeAppThemeTip: Tip {
    // rest of the tip implementation
    @Parameter
    static var isPremiumUser: Bool = false
}
```

Expanding the macro reveals a simple implementation:

```
static var $isPremiumUser: Tips.Parameter<Bool> =
  Tips.Parameter(Self.self, "+isPremiumUser", false)
  {
    get {
            $isPremiumUser.wrappedValue
    }
    set {
            $isPremiumUser.wrappedValue = newValue
    }
}
```

Let's delve into the macro implementation. Since TipKit wants to work with a generic type, the macro creates another variable called `$isPremiumUser` of the `Tips.Parameter` type (based on `Bool`) and a default value of `false` (as initially defined in our static variable).

The macro also creates a **getter** and a **setter** so our tip can respond to app state changes.

However, the macro handles another thing that helps us: making our parameter value **persistent**. In this case, the answer to the question "Is the user premium?" is probably persistent anyway. However, there are cases where it's not that obvious. For example, feature usage tracking is not normally persistent.

Now that we have a parameter let's add our first display rule.

Defining our display rules

Are we defining display "rules", in plural?

Yes! TipKit supports multiple display rules to support more complex situations. However, first, let's start with one tip:

```
struct ChangeAppThemeTip: Tip {
    @Parameter
    static var isPremiumUser: Bool = false

    var rules: [Rule] {
        [
            #Rule(Self.$isPremiumUser) {
                $0 == true
            }
        ]
    }
}
```

In this code example, we use a macro to build a data type called `Rule` that contains a predicate expression. That predicate expression compares the given type to a specific value.

In this case, we compare the `$isPremiumUser` value to `true`.

Now, let's go back to the `rules` variable. We can add more rules that support our tip display logic. TipKit performs an AND operator between the different tips, and if the result is true, then the tip is displayed (unless the user or the app dismisses it, obviously).

How can we modify the value the rule is based on? Let's see.

Connecting the parameter to our app logic

We need to connect the tip parameters to our app logic to finalize our work. Notice that the parameter is a static variable. This means that we can modify it from anywhere in our app, even if we don't have a reference for the tip instance.

Let's see an essential parameter modification:

```
let tip = ChangeAppThemeTip()
var body: some View {
    VStack {
        Button("Change isPremium parameter") {
            ChangeAppThemeTip.isPremiumUser.toggle()
        }
        TipView(tip)
    }
}
```

This code example shows a basic UI with a button that toggles the static `isPremiumUser` variable, which we created in our tip earlier. Toggling that value also shows and hides the `TipView` view down the VStack.

However, adding a button that toggles a tip is not a real-world example of using a rule parameter. A more practical example would be connecting it directly to the user's premium state using a `Combine` stream – something like the following code:

```
let premiumManager = PremiumPurchaseManager()
let premiumStatusSubscription =
  premiumManager.premiumPurchasePublisher
    .assign(to: \.isPremiumUser, on:
      ChangeAppThemeTip.self)
```

In this code example, we have a premium purchase publisher and we assign its output to our tip's `isPremiumUser` parameter. This is a more elegant way to link the rule logic to our app.

Now let's discuss the other type of rules – events.

Adding a rule based on events

When we display a tip based on a state, it's usually only displayed when the user can use a particular feature. However, there are cases when we want to display a tip when we think the user is ready to take our app to the following usage level.

For example, if we create a music app and the user adds a few songs, maybe it's a good idea to tell them about making a playlist. Or, if we are working on a dating app, maybe it is worth suggesting modifying the search filter if the user hasn't chosen any of the profiles viewed.

For these types of tips, we can create a rule based on tracking events. The idea is to define an event representing the user's relevant action. For example, I can add a task, view a profile, and more. Afterward, we create a rule based on the number of events tracked within a time frame or generally.

Let's see a code example for a tip suggesting the user add a list of to-dos. We'll start by defining our tip:

```
struct AddListTip: Tip {

    static let didAddATaskEvent = Event(id:
      "didAddATaskEvent")

    var rules: [Rule] {
        #Rule(Self.didAddATaskEvent) {
            $0.donations.count > 3
        }
    }
}
```

The tip goal is to suggest the user add to a list of to-dos. We create an event called `didAppTaskEvent` that helps us track the number of times the user adds a new to-do.

The second thing we do here is to create a new rule that returns `true` if the number of tracked events exceeds three.

This is a different rule constructor that handles event tracking instead of a state.

The last piece of the puzzle shows the tip and track of an event:

```
struct EventRuleTipExample: View {

    let tip = AddListTip()
    @State var todos: [Todo] = []

    var body: some View {
        VStack {
            TipView(tip)
            List(todos) { todo in
                Text(todo.title)
            }
            Spacer()
            Button("Add task") {
                todos.append(Todo(title: "New Task"))
                Task{ await
                    AddListTip.didAddATaskEvent.donate()}
            }
        }
```

```
        }
}
```

The event tracking operation is referred to as `donate()`, while the total number of tracked events is known as **donations**.

We can also check for events tracked in a specific time range:

```
$0.donations.donatedWithin(.days(3)).count > 3
$0.donations.donatedWithin(.week).count < 3
```

This example checks whether the number of events exceeds three in the last three days or one week.

Now, it's important to distinguish between the number of events tracked and just checking the database for the number of to-dos.

We could easily check the user's number of to-dos in their database and change that to a state-based rule. However, this solves a different use case – not the number of times the user added a task with the app, but rather the number of tasks the user has in general.

Grouping tips with TipGroup

When our app becomes more extensive and feature-rich, handling a large set of tips can become cumbersome. Trying to coordinate all these tips using rules can lead to a situation wherein tips appear outside the intended order and even together.

To address that, we can use the `TipGroup` class to group tips and present them individually in a particular order.

Let's see an example for a `TipGroup` class usage:

```
@State var tips = TipGroup(.ordered) {
        FirstTip()
        SecondTip()
    }

    var body: some View {
        Button("Settings") {

        }.popoverTip(tips.currentTip)
    }
```

In this example, we created a state variable called `tips` of the TipGroup type. We passed `.ordered` for its priority parameter and added two tips using its builder. In the code itself, we attached our `TipGroup` instance to a button using the `popoverTip` view modifier, passing the group's current tip.

Using the `.ordered` parameter ensures that the tips will appear in the order in which we added them to the builder. TipKit will show the next tip once all the previous tips have been invalidated.

The other parameter we can use is `firstAvailable`, which shows the next tip that is eligible for display.

Grouping tips together can help manage a large collection of tips in our project. However, looking at the code example again, we can see that there might be a problem with the way we implemented the TipGroup in the view. Imagine we have a TipGroup with a `SettingsTip` type and a `ProfileTip` type. When using the TipGroup for settings and profile buttons, we can't control which tip appears where.

To solve that, we can cast the `currentTip` variable to the desired tip type. Let's see that in the following code:

```
@State var tips = TipGroup(.ordered) {
    SettingsTip()
    ProfileTip()
}

var body: some View {
    Button("Settings") {

    }.popoverTip(tips.currentTip as? SettingsTip)

    Button("Profile") {

    }.popoverTip(tips.currentTip as? ProfileTip)
}
```

In this code example, we have a TipGroup with two tips – for the settings button and for the profile button.

When we use the `popoverTip` view builder, we cast the `currentTip` instance to the corresponding type according to the button. This technique takes advantage of how the `popoverTip` signature looks:

```
public func popoverTip(_ tip: (any Tip)?...)
```

Since `popoverTip` accepts `nil` as an argument, we can ensure that only relevant tips will appear from the TipGroup.

Rules are only one aspect of defining the appearance logic. Another crucial element is determining its frequency. Let's see how to customize that as well.

Customizing display frequency

In the previous section, we discussed creating display logic for our tips using rules and tip groups. However, tips can overwhelm users; there's a fine line between helping the user and disturbing them. Adjusting all the rules to set a reasonable limit on the number of tips the user sees can be challenging. For that problem, we can manage the frequency at which our tips display.

Let's start with setting the max display count for a tip.

Setting the max display count for a specific tip

The first and essential thing we can do is set the maximum number of a specific tip type that can be displayed.

We do that by adding a new variable to our tip called `options`:

```
struct AddListTip: Tip {
    var options: [TipOption] {
        Tips.MaxDisplayCount(2)
    }
}
```

In this code example, we use the `MaxDisplayCount` static function of the `Tips` namespace. That definition means that the tip will be displayed a maximum of two times, and afterward, it will be invalidated, overriding the rest of the rule's logic. That's a great way to ensure that a specific tip doesn't overwhelm users.

However, there's another excellent way to ensure a calmer user experience: display frequency.

Setting our tips' display frequency

We just learned how to limit a particular tip to a certain number of appearances. Another way to handle tip appearance is to define its frequency.

Let's look at the following code:

```
struct MyApp: App {
    init() {
        try? Tips.configure([.displayFrequency(.daily)])
    }
}
```

The code example shows how we can limit the total number of tips displayed to one per day.

The `.displayFrequency(.daily)` expression means that TipKit will show no more than one tip per day. Obviously, we have additional frequency options: hourly, weekly, monthly, and immediate.

We can configure specific tips to ignore the system display frequency:

```
struct AddListTip: Tip {
    var options: [TipOption] {
        Tips.IgnoresDisplayFrequency(true)
    }
}
```

In this code example, the `AddListTip` tip ignores the system definition for general display frequency.

Setting the max display count for a specific tip and defining a display frequency for all tips is a great way to fine-tune the user's tips experience.

Summary

In this chapter, we discussed the importance of TipKit, added our first tip, customized its design and behavior, learned how to manage tips better by grouping them, and minimized their appearance by setting their display frequency. By now, we are fully prepared to implement TipKit in our apps.

TipKit touches on a severe app aspect: engagement and feature exploration. It looks like it supports many product requirements!

In the next chapter, we'll discuss how to work seamlessly with one of our most important data sources: the network.

8
Connecting and Fetching Data from the Network

Finding an app that isn't connected to a server is extremely difficult. Most apps don't operate alone—they need to authenticate their users, fetch information, and allow their users to perform actions that eventually will be synced back to the server.

Due to this, it is important to understand how networking works—not how HTTP works in general, but how iOS apps work with the server efficiently and simply.

In this chapter, we will cover the following topics:

- Understanding mobile networking
- Handling HTTP requests, including their responses
- Integrating network calls within app flows
- Exploring how Combine works with networking

Let's start understanding how the network fits into our app architecture.

Technical requirements

For this chapter, you must download Xcode version 15.0 or above from Apple's App Store. Search for Xcode in the App Store and select and download the latest version. Launch Xcode and follow any additional installation instructions that your system may prompt you with. Once Xcode has fully launched, you're ready to go.

You'll need to run the latest version of macOS (Ventura or above).

You can also download the sample code from the following GitHub link: https://github.com/PacktPublishing/Mastering-iOS-18-Development/tree/main/Chapter%208

Understanding mobile networking

Working with the network is a crucial part of developing apps for iOS. Understanding how the network fits into our app architecture is essential, but what does it mean? Watching simple tutorials about performing a URLSession request is fine, but real-world apps don't work that way.

Before we dive any deeper, let's recap what a basic app architecture looks like:

- **UI layer**: This is responsible for presenting UI to the user, including responding to user inputs. The UI layer consists of the SwiftUI/UIKit views and view models.
- **Business logic**: This is responsible for manipulating data while managing the basic application logic.
- **Data layer**: This is responsible for storing and retrieving data entities related to the business logic.

I guess I'm not surprising you here with this three-layer architecture, as most mobile apps work in a similar architecture.

When we begin to understand where the network job fits, we will have to look at the data layer and, in some specific cases, the business logic layer (for example, when working with analytics or third-party libraries). However, why do we need to look at the layers? To understand why our network activity is relevant for the data layer, let's go over our main network goals:

- **Syncing information** to and from our backend
- Handling **authentication**
- Logical activity that requires a server

In most apps, networking is needed to sync data with our backend. The data layer functions as the primary repository of truth for entities. Attempting to access entities directly from the network in other layers will undermine this principle.

Figure 8.1 shows a basic app architecture and the network location:

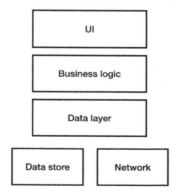

Figure 8.1: Basic app architecture

In *Figure 8.1*, we can see that the network is one of the components that build up the data layer. The basic idea of most apps is for the network to become a data source and fill the app's data store.

For example, in a music app, the network layer might reach out to the backend, fetch albums and songs, and store them in local storage such as **Core Data**. The network layer is also built upon different components to function correctly.

We can think of network operations as a factory production line. We request a piece of information and take care of the returned data package, transferring it through several stages until we properly store it in our local store or present it.

Before we review the stages a data package can undergo, let's try to build a network request together. We'll start by reviewing the basic HTTP request methods.

Handling an HTTP request

An HTTP request is a message the client sends to a server to request information and perform an action. The server processes that request and returns a response to the client. Clients indeed perform HTTP requests asynchronously to leave the main thread free. However, the connection between the client and server is **synchronous** as the client waits for the server's response to complete the request operation.

The primary HTTP request component is the request method, which indicates the request's main goal. Let's go over some of the basic HTTP methods now.

Basic HTTP request methods

The REST API is based on the idea of a request-response style, and it's a one-directional communication with our backend. The REST API has eight methods to use when communicating with the backend. However, in most cases, we will use the following four methods:

- GET: This is used to retrieve information only from the server. It should be a safe call, meaning that performing a GET request shouldn't affect the backend data.
- POST: The POST method is often used to submit data to the backend. In many cases, the POST method performs changes in the backend data store or changes a user state.
- PUT: We use PUT to create or update objects. Unlike the POST method, PUT is considered idempotent. We can send multiple identical PUT requests and expect the same effect as sending one request.
- DELETE: As the name states, we use DELETE to delete objects. Obviously, we can use POST to do that, but with DELETE, we are aligned with the standards.

It is worth mentioning that, technically, we can even use GET to make changes to the server. However, the proper method ensures predictability and reliability and is aligned with the REST principles.

To perform a basic HTTP request, we should first be familiar with the `URLSession` class.

Working with URLSession

We can use a class called `URLSession` to perform and manage network requests. `URLSession` is part of what Apple calls the **URL Loading System**, which in turn is part of the **Foundation** framework.

The `URLSession` class is responsible for coordinating different URL requests in our app. Let's see how to perform a basic `GET` response using `URLSession`:

```
let urlString =
  "https://jsonplaceholder.typicode.com/posts"
if let url = URL(string: urlString) {
   var request = URLRequest(url: url)
   request.httpMethod = "GET"
   let session = URLSession(configuration: .default)
   let task = session.dataTask(with: request) { (data,
     response, error) in
   }
   task.resume()
}
```

This code example creates an object called `URLRequest` based on a particular URL. The `URLRequest` class encapsulates the information we need to perform a specific URL request. It usually consists of the following information:

- The request base URL
- The request method – `GET`, `POST`, `PUT`, or `DELETE`
- The request HTTP headers

Notice that the `URLRequest` structure doesn't perform the actual HTTP request or contain its response information. The `URLSession` class is responsible for conducting and managing the different HTTP requests.

There are two ways to initialize a `URLSession` instance:

- We can call the static `shared` property and use it as a **singleton**. We do that if we want to simplify our implementation without needing to customize how we handle requests or have different requirements for different areas in the app:

    ```
    let session = URLSession.shared
    ```

- If we need more flexibility, we can create an instance of `URLSession` (like in the last code example) and initialize it with our own configuration.

A configuration object allows us to fine-tune our requests better. For example, we can define each request as containing additional headers, setting the timeout and caching, or even cookie acceptance policies.

Here's a code example for setting up a `URLSession` class with a specific timeout duration and no caching:

```
let configuration = URLSessionConfiguration.default
configuration.timeoutIntervalForRequest = 10
configuration.requestCachePolicy =
  .reloadIgnoringLocalCacheData
let session = URLSession(configuration: configuration)
```

In this code example, we created a configuration object, set its `timeoutIntervalForRequest` value to `10`, and defined the cache policy to be ignored.

When we work with a shared `URLSession` object, there's no way to customize its configuration, and it will use the default one.

Now that we know how to perform a basic `GET` or `POST` request, let's see what we can do with the response.

Handling the response

The request response is handled using three stages: error handling, serialization, and data storage. We need to handle each one of the stages carefully and even consider having a dedicated class or function to simplify the process and separate the concerns.

As mentioned, the first stage is error handling. Let's discuss it, as it is a crucial part of networking.

Implementing error handling

I believe error handling wouldn't get a whole section in many frameworks. It is usually a straightforward topic: we perform a task, something goes wrong, and we receive an error.

However, with networking, we are working in a volatile environment where many things have the potential to fail the process.

Here's a partial list of things that can go wrong:

- There is no network
- There's a network, but the device cannot reach the internet
- The device can reach the internet but with a very slow connection
- We have a stable connection, but the request cannot reach the backend
- The request found the backend, but it didn't respond

The error list can go on and on, ranging from network issues to security to server errors.

To simplify the idea, we can divide the errors into two main groups: network-related issues and server-side problems.

To understand the difference between network and server-related issues, let's have another look at how we created a data task:

```
let task = session.dataTask(with: request) { (data,
   response, error)
```

We can see that the data task response contains three parameters – `data`, `response`, and `error`.

Network-related errors are part of the `error` object, and server-related errors are mostly part of the `response` object and sometimes even part of the `data` object.

To handle a network error, we should look into `URLError`:

```
if let error = error as? URLError {
   switch error.code {
   case .cannotFindHost:
      // notify the user.
   default:
      print("Error: \(error)")
   }
   return
}
```

In this code example, we performed a switch statement to understand our network error. In this case, we decided to handle one use case of `cannotFindHost`. However, there are at least 20 different error codes we can handle. To read the full and updated list, we should look at Apple documentation at `https://developer.apple.com/documentation/foundation/urlerror`.

Unlike network-related errors, server-related errors are more complex. First, we are dependent on another partner—our server. How the server implements its error-handling logic significantly influences how we handle it in our app.

Let's understand that by examining the server response:

```
if let httpResponse = response as? HTTPURLResponse {
        switch httpResponse.statusCode {
        case 200..<300:
            print("Success:
              \(httpResponse.statusCode)")
        case 400..<500:
            print("Client Error:
              \(httpResponse.statusCode)")
```

```
            case 500..<600:
                print("Server Error:
                   \(httpResponse.statusCode)")
            default:
                print("Other Status Code:
                   \(httpResponse.statusCode)")
            }
        } else {
            print("Invalid HTTP Response")
        }
```

We first cast the response into the HTTPURLResponse type, representing a general URL response.

The response includes a status code, which the server sends back to us. In most cases, the code will be part of the following three groups:

- 200..299: The server successfully responded to our request
- 400..499: The server returns an error due to a bad client request
- 500..599: The server returned an error due to an internal server error

In short, there are three cases – everything went well, it is the client's fault, or it is the server's fault.

However, in real life, things are more complex. Sometimes, the server returns a response code of 200 (success) but includes an error in the response data. There is nothing wrong with doing that – the server can choose how to handle problems. It's our responsibility to parse the response correctly.

If we need to parse the response ourselves to extract the error, it is better to create a function that receives the data, response, and error parameters and throws an error in case it finds one:

```
func handleResponse(data: Data?, response: URLResponse?, error: Error?) throws {
    if let error = error {
        throw error
    }
    
    guard let httpResponse = response as? HTTPURLResponse
      else {
        throw NetworkingError.invalidResponse
    }
    
    switch httpResponse.statusCode {
    case 200..<300:
        if let responseData = data {
            if let errorData = try?
              JSONDecoder().decode(ErrorResponse.self,
```

```
                    from: responseData) {
                    throw NetworkingError.dataError
                }
            }
        case 400..<500:
            throw NetworkingError.clientError(statusCode:
              httpResponse.statusCode)
        case 500..<600:
            throw NetworkingError.serverError(statusCode:
              httpResponse.statusCode)
        default:
            throw NetworkingError.otherError
        }
    }
```

This long `handleResponse` function does precisely what we've discussed. In case of a successful response, it checks the error object, the response code, and the data itself.

To use that function, we need to call it within the response closure:

```
    let task = session.dataTask(with: request) { (data,
      response, error) in
        do {
            try handleResponse(data: data, response: response,
              error: error)
        } catch let error {
            print("Error: \(error)")
        }
    }
```

The great thing about the `handleResponse` function is that we can ensure that we can continue handling the response data after the `try` statement because we have dealt with any error.

If you look again at the `handleResponse` function, you'll see that we decode the response to look for an error.

Deserializing the response is a major step in handling a network response. Let's discuss it a little bit further.

Deserializing a network response

In most apps, the response we get from the server is based on JSON data structure. JSON is an industry standard for delivering network responses along with XML.

Swift has built-in support for parsing JSON structures into Swift structures, using tools such as the `Codable` protocol and `JSONDecoder` classes.

In theory, it sounds perfect—all we need to do is decode our response to a data model. However, there are more factors we need to consider:

- **Supporting general responses**: Not all responses are data models. There are network responses that include general messages. For one, in our `handleResponse` function example, we saw a response that may have contained an error message. This means that when we think about our data models, general network responses should be among them.
- **Assuming there's always an object array**: Decoding a single object is straightforward, but in many cases, we also need to handle decoding an array of objects. That sounds trivial, but supporting both formats can be a hassle. To simplify the decoding process, it is better to always support an array of objects, which is a decision that we need to coordinate with our backend developers.
- **Mixed structures**: A response can contain different model types and even nested data structures. This is not always trivial, so our data structures must be more dynamic and modular to support various responses.
- **Model transformations**: Our local app models are structured to be efficient and convenient to use with the business logic and UI layers. However, who said that the backend response structure is aligned with what is suitable for our app? This means we must transform the response data model to our local data model.

Deserializing data models is indeed a complex task, and trying to match our data models to the response structure we receive from our backend is only sometimes the best idea. Remember that our data models must suit our app needs and not necessarily follow the backend methodology.

Let's take a simple JSON received from the server:

```
{
  "id": 1,
  "name": "John Doe",
  "email": "john@example.com"
}
```

That's a contact structure. However, we want to use different names in our app so we can use the `CodingKey` protocol to ensure we perform the transformation correctly:

```
struct Contact: Codable {
    let id: Int
    let fullName: String
    let userEmail: String
    
    // Define custom coding keys to match JSON keys
    private enum CodingKeys: String, CodingKey {
        case id
        case fullName = "name"
```

```
        case userEmail = "email"
    }
}
```

Decoding the server response using the `Contact` structure now becomes much simpler:

```
let errorData = try? JSONDecoder().decode(Contact.self,
    from: responseData)
```

In this example, we map the `name` value to `fullName` and `email` to `userEmail`. We decode it using the `JSONDecoder` class. Understanding the `CodingKey` protocol is a crucial key to decoding server responses.

There are cases where the whole structure of the server response is entirely different than our data models, and in those cases, we need to create a dedicated structure to parse the response. However, sometimes, we can still use our data model as part of the structure. Let's look at the following example:

```
struct ServerResponse: Codable {
    let responseID: String
    let timestamp: String
    let orgID: String
    let contact: Contact
}

let jsonString = """
{
    "responseID": "12345",
    "timestamp": "2024-03-25T12:00:00Z",
    "orgID": "5678",
    "contact": {
        "id": 1,
        "fullName": "John Doe",
        "userEmail": "john@example.com"
    }
}
"""
let jsonData = jsonString.data(using: .utf8)!
let response = try
    JSONDecoder().decode(ServerResponse.self, from: jsonData)
```

In this code example, the server returns additional information besides the contact object. So, we can create a dedicated data structure for the response—`ServerResponse` (in this case). In addition to general information, the `ServerResponse` struct contains the `Contact` struct. So, we can see a modular approach here—we can parse our server response using `Codable` and still use our data model objects to receive the information.

The next step is to store our data model in our data store.

Building a data store

A disclaimer: not every network call requires us to store the results in a data store. For instance, authentication and logic calls have different goals. However, this chapter will focus mainly on data-related calls responsible for building our local data store.

That leads us to our next point: what is the role of the data store?

So, a data store is a structured mechanism for managing and storing data that serves the application's main business logic and UI.

Unlike many online examples, the application business logic usually doesn't work directly with the network responses – these need to be adjusted and saved in our store, which acts as the UI data source.

Let's look at *Figure 8.2*:

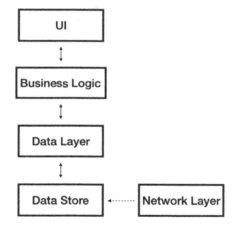

Figure 8.2: Working with the datastore

Figure 8.2 shows how the data layer works directly with the data store, as the network layer fills the data store with more information.

The data store doesn't have to be persistent—that's an engineering decision. However, in most cases, it is a structured store. A structured store has pre-defined models, relations between entities, and often even query capabilities. These characteristics distinguish the data store from simply caching the network responses.

To follow the separation of concerns principle, it is better to have dedicated classes to handle each step of the process.

First, we'll create a `DataStore` class:

```swift
class DataStore {
    private var contacts: [Contact] = []
    
    func updateContacts(with newContacts: [Contact]) {
        contacts = newContacts
    }
    
    func getAllContacts() -> [Contact] {
        return contacts
    }
}
```

The `DataStore` class not only stores the data but also has methods that help store and retrieve entities.

Assuming we already have a network handler from the previous examples, we are now going to create a sync class that coordinates the process of fetching the data and storing it:

```swift
class SyncManager {
    private let dataStore: DataStore
    
    init(dataStore: DataStore) {
        self.dataStore = dataStore
    }
    
    func syncData() {
        NetworkHandler.fetchData { result in
            switch result {
            case .success(let data):
                do {
                    let contacts = try JSONDecoder().decode([Contact].self, from: data)
                    self.dataStore.updateContacts(with: contacts)
                    print("Data synced successfully")
                } catch {
                    print("Error decoding data:", error)
                }
            case .failure(let error):
                print("Error fetching data:", error)
            }
        }
    }
}
```

The `SyncManager` class uses the `NetworkHandler` class to fetch the information from our backend, parses the results into `Contact` entities, and stores them in our data store. Using this design pattern, we can easily replace the data store implementation to be persistent without modifying the other classes.

Now that we have a data store, let's try to understand how to make our app more efficient.

Integrating network calls within app flows

We already know how to perform a network call, parse it to data objects, and create a data store. We also know how to handle errors, and we learned that it's important to separate the concerns into different classes and components.

However, it feels like a technical discussion. Performing a URL connection in iOS is one of the most basic tasks. Let's try to upgrade our discussion and discuss methodology.

First, we should think of streaming data from the network as an atomic task in our app's data synchronization mechanism. It's up to us to decide when to call our server for more data. From our discussions, it looks like we need to contact the server just before we want to display the information, but it doesn't have to be like that.

Let's discuss the different strategies we can use when working with our backend. We'll start with the **just-in-time** fetching technique.

Just-in-time fetching

The just-in-time fetching technique is very common and also the simplest one. With it, we don't present anything on the screen before we get a response from the server. Instead, we show a loader indicating that we are fetching data.

In just-in-time fetching, we don't preserve the information in a data store; instead, we store the information in the view state or the view model. Here's a simple example of just-in-time fetching:

```swift
import SwiftUI

struct ContactsView: View {
    @State private var contacts: [Contact] = []
    @State private var isLoading = false

    var body: some View {
        NavigationView {
            List(contacts) { contact in
                VStack(alignment: .leading) {
                    Text(contact.name).font(.headline)
                    Text(contact.phoneNumber).font(.subheadline)
```

```
            }
        }
        .navigationTitle("Contacts")
        .onAppear {
            fetchContacts()
        }
        .overlay {
            if isLoading {
                ProgressView("Loading...")
            }
        }
    }
}

private func fetchContacts() {
    isLoading = true
    NetworkHandler().fetchData { fetchedContacts in
        contacts = fetchedContacts
        isLoading = false
    }
}
}
```

In this code example, we have a list that is based on the state variable of contacts. When the view appears, we call the `fetchContacts` method to fetch the list of contacts and, in the meantime, show a loading message.

Besides its simplicity, the just-in-time technique is great for apps that must ensure that the data they display is up to date, such as financial apps or live sports scores. The downside here is that we provide a poor user experience and depend on the network state.

If possible, we should pick a slightly better technique, often called **read-through cache**.

Read-through cache

The read-through cache technique is also a popular way to present data to the user, even though most developers are unaware of its name.

Using the read-through cache approach, we display our local data to the user while going to our backend to refresh our data.

Let's see a code example for that:

```swift
import SwiftUI

struct ContactsView: View {
    @State private var contacts: [Contact] = []
    
    var body: some View {
        NavigationView {
            List(contacts) { contact in
                VStack(alignment: .leading) {
                    Text(contact.name).font(.headline)
                    Text(contact.phoneNumber).font(.subheadline)
                }
            }
            .navigationTitle("Contacts")
            .onAppear {
                loadContacts()
            }
        }
    }
    
    private func loadContacts() {
        contacts = loadFromCache()
        
        NetworkHandler().fetchData { fetchedContacts in
            contacts = fetchedContacts
            saveToCache(contacts: fetchedContacts)
        }
    }
}
```

In this code example, we load the contacts from the cache (or from the local store) when the screen appears and then go to the network to refresh our data set. The read-through cache technique is great when quick access to data is crucial because it is not up-to-date, for example, in news or e-commerce apps.

You've probably noticed that both the just-in-time and read-through cache techniques require us to load the page information fully from the backend, regardless of the amount of information we have.

Now, what if we know upfront that we have a huge number of records to fetch, so big that it can even cause our request to time out? In this case, we can choose the **incremental loading** technique.

Incremental loading

There are cases wherein we can expect to fetch a vast number of records. A social feed, for instance, can have an infinite number of posts. Well, it's not really infinite, but we can relate to that number as infinity.

When the number is too big to fetch in one request, we can use incremental loading.

With incremental loading, we fetch a set of records each time with each request and store the last record index for the next time.

Here's an example of incremental loading:

```swift
class IncrementalLoader {
    var currentPage = 1
    let itemsPerPage = 10
    var contacts = [Contact]()
    
    func loadNextPage() {
        guard let url = URL(string: "https://api.example.com/contacts?page=\(currentPage)&limit=\(itemsPerPage)") else {
            print("Invalid URL")
            return
        }
        
        let task = URLSession.shared.dataTask(with: url) { [weak self] (data, response, error) in
            guard let self = self else { return }
            
            
            do {
                let newContacts = try JSONDecoder().decode([Contact].self, from: data)
                DispatchQueue.main.async {
                    self.contacts.append(contentsOf: newContacts)
                    print("Fetched Contacts: \(newContacts)")
                    self.currentPage += 1
                }
            } catch {
                print("Error decoding JSON: \(error)")
            }
        }
        task.resume()
    }
}
```

In this example, we have a class named `IncrementalLoading`, which is responsible for loading the next set of records with the function named `loadNextPage`. Incremental loading is also called **pagination loading** because it mimics the concept of paging through a book. In our `IncrementalLoading` example, we have an index that points to the last record index fetched, and a variable named `itemsPerPage` that defines how many items to fetch on each page.

While incremental loading solves handling a large amount of data, there are several factors we need to consider:

- **Complexity**: Incremental loading is considered a relatively complex design pattern, mainly because we build our data collection in stages according to user interaction. For example, one classic way to implement incremental loading is by having a SwiftUI `List` view or a UIKit `TableView` view. In these cases, we would like to fetch the next set of records when the user reaches the bottom of the list. However, things can become complex when we allow the user to edit or delete records since that can affect the index variable.

- **Memory consumption**: It's true that incremental loading is optimized to handle a significant amount of information. However, we are still talking about storing a large amount of information in our memory. While the user is paging through our data, our local data store can become bigger, mainly if it contains rich media such as images and videos. It is essential to have a mechanism that can release records in case of high memory usage.

- **Contextual relevance**: We need to remember that our chosen design pattern needs to support a specific product need. Incremental loading is relevant in cases wherein we don't need all the data at once. Social feeds or search results are great examples of data that can be browsed chunk by chunk. However, in cases where the user requires immediate access to all the data, such as in data analysis, incremental loading might not be suitable.

Considering the different factors mentioned, we understand that, similar to many design patterns in computer science, incremental loading presents a tradeoff between different aspects such as performance, complexity, experience, and more. It's up to us to choose the right design pattern that fits our needs.

The three design patterns we discussed now require different endpoints for different types of data and other screens, which sounds logical. However, there's another way to handle data that changes over time and still provides an amazing experience to the user – delta updates.

Full data sync with delta updates

Before we discuss full data sync with the delta updates method, let's talk about problems that we have with multiple endpoints:

- **Efficient network calls**: The need to request the same data repeatedly, even if nothing has changed, seems inefficient. We can use the cache to present previous results, but that only solves performance issues. We still need to perform the same request to understand whether there are updates.

- **Incomplete database**: Each endpoint retrieves different data and sometimes different entities. We know that in many cases, the entities are related (such as to-one and to-many relationships), and having multiple endpoints to fetch them probably means our data won't be complete. That seems acceptable – we're focused on a mobile app and not a server. However, having an incomplete data store can result in a poor experience. Users may encounter updated information on one screen, navigate to another, and view outdated data while waiting for the screen to refresh from the server. If both screens contain related data, it can result in a poor experience.
- **App performance**: We often believe that performance is only about CPU and Swift code efficiency. However, our devices are strong enough to handle most tasks without a hiccup. In contrast, network requests cause users to wait even if they have the latest hardware. Having a network call on each screen greatly impacts the user experience.

Delta updates are a solution that can handle some of the problems we described with endpoints in the previous section. With delta updates, we fetch all the information at the app's initial launch and, from this point, retrieve only the changes.

We do that by storing a bookmark representing our data's last updated timestamp. When we ask the server, "Do you have any updates for me?", we send this bookmark, get the new changes (if any), receive a new bookmark, and store it.

Here's a code example for contacts delta sync. We start with the `syncContacts` function:

```
class ContactsSyncManager {
    
    let userDefaults = UserDefaults.standard
    let lastUpdatedKey = "lastUpdatedTime"
    let syncEndpoint = URL(string:
      "https://example.com/api/sync/contacts")!
    
    func syncContacts() {
        var request = URLRequest(url: syncEndpoint)
        request.httpMethod = "POST"
        request.addValue("application/json",
          forHTTPHeaderField: "Content-Type")
        
        let lastUpdatedTime = userDefaults.double(forKey:
          lastUpdatedKey)
        let requestBody = ["lastUpdatedTime":
          lastUpdatedTime]
        request.httpBody = try?
          JSONSerialization.data(withJSONObject:
          requestBody)
```

```
            URLSession.shared.dataTask(with: request) { [weak
              self] data, response, error in
                self?.processDeltaUpdates(response: response)
            }.resume()
        }
```

The code example does exactly what we described earlier—it saves a bookmark called `lastUpdatedDate`. Initially, we fetch all the data and save the new `lastUpdatedDate` value we get from the server. The next time we perform the sync operation, we get only the changes. Now, let's implement the `processDeltaUpdates` function:

```
        private func processDeltaUpdates(response:
          ContactsDeltaUpdateResponse) {
            // Here you can handle the new, deleted, and updated contacts
  as needed
            print("New Contacts:
              \(response.newContacts.count)")
            print("Deleted Contacts:
              \(response.deletedContacts.count)")
            print("Updated Contacts:
              \(response.updatedContacts.count)")

            userDefaults.set(response.lastUpdated, forKey:
              lastUpdatedKey)
        }
    }
```

The `processDeltaUpdates` function receives a response that contains only the changes that have occurred in the server since the last sync.

That's why the response is structured into three groups: deleted, new, and updated. With each one, we need to handle the data differently.

Some critical notes we need to consider here are as follows:

- **Extremally large data sets**: The delta updates pattern is not relevant for very large data sets. For example, a social app feed can have millions of records, and fetching all of them from the start is impossible. For that issue, we can use pagination.
- **The initial loading can be long**: Since we fetch all the data at the beginning, we need to deliver a corresponding user experience.

- **Deleted items**: Syncing deleted items is always a crucial topic. We need to actively delete items that no longer exist, so the response from the server should contain items we need to delete.
- **Sync triggers**: Since we perform the sync operation at the beginning, it looks like it's the only time we should do that. However, there are more occasions when we need to refresh our data. For example, when we perform data changes such as calling the server to add a new item or receiving a push notification, we should think about the different cases when something can change in our server during the app runtime and try to refresh our data.

It's important to understand that none of the solutions are perfect. Sometimes, it is a good idea to combine different approaches—for example, use delta sync in general, but maybe use pagination for a specific screen.

We should consider the different approaches as a toolbox with several tools, each suitable for various problems or data structures.

Now that we understand how to handle requests and use different patterns to incorporate the calls in our app flows, let's see another way to handle networking in iOS.

Exploring Networking and Combine

Networking is a great place to start if you haven't worked with Combine. Combine is a framework that declaratively handles a stream of values over time while supporting asynchronous operations.

Based on that description, it looks like Combine was made for networking operations!

In this chapter, we are not going to discuss what Combine is – for that, we've got *Chapter 11*. However, we are going to discuss it now because Combine is a great way to solve many networking operations problems.

Since Combine is built upon publishers and operators, it is simple to create new publishers that stream data from the network.

Let's try to request the list of contacts from previous examples using a Combine stream. We'll start with creating a publisher that performs data fetching from the network and publish a list of contacts:

```
class ContactRequest {
    func fetchData() -> AnyPublisher<[Contact], Error> {
        let url = URL(string:
          "https://api.example.com/contacts")!

        return URLSession.shared.dataTaskPublisher(for:
          url)
            .map { $0.data }
            .decode(type: [Contact].self, decoder:
              JSONDecoder())
```

```
            .eraseToAnyPublisher()
    }
}
```

The publisher utilizes URLSession's `dataTaskPublisher` method to execute the network request and publish the retrieved data. We then extract the data using the map operation and decode it into a list of `Contact` items. If something goes wrong, the publisher will report an Error. We wrap this function in a class named `ContactRequest` to maintain separation.

Now, let's create a small `DataStore` class so we can store the results and publish them:

```
class DataStore {
    @Published var contacts: [Contact] = []
}
```

The `@Published` property wrapper creates a publisher for contacts so that we can observe the changes easily.

Now, we can use the `fetchData()` function to read the results and store them:

```
class ContactsSync {
    let contactRequest = ContactRequest()
    let dataStore = DataStore()

    func syncContacts() {
        contactRequest.fetchData()
            .sink(receiveCompletion: { completion in
                switch completion {
                case .finished:
                    print("Data fetch completed
                        successfully")
                case .failure(let error):
                    print("Error fetching data: \(error)")
                }
            }, receiveValue: { [weak self] contacts in
                self?.dataStore.contacts = contacts
            })
            .store(in: &cancellables)
    }

    private var cancellables = Set<AnyCancellable>()
}

let contactsSync = ContactsSync()
contactsSync.syncContacts()
```

The `ContactsSync` job is to fetch contacts using the `ContactRequest` class and to store them in the data store using the `DataStore` class.

The Combine example has several advantages:

- **Clear and consistent interface**: The publisher interface is consistent and known. It is always built from data/void and an optional error. New developers don't need to learn and understand how to read/use it.
- **Built-in error handling**: Not only do we have a consistent interface that also contains errors, but also, when one of the stages encounters an error, it interrupts the flow and channels it downstream. We have already seen that error handling is a critical topic in networking in many cases.
- **Asynchronous operations support**: We often think that a network operation contains one asynchronous operation: the request itself. However, many steps in the stream can be asynchronous – including preparing the request by reading local data, processing the response, and storing the data at the end of the stream. Combine streams are perfect for performing all those steps asynchronously.
- **Modularity**: The capability of building a modular code is reserved not only for the Combine framework, but the custom publishers and the different operators make Combine streams a joyful framework to implement when dealing with networking. Remember that we said that networking is like a production line (under the *Understanding mobile networking* section)? So, Combine makes it easier to insert more steps into the stream; some of them are even built into the framework.

Adding reactive methods to our code doesn't mean we need to discard all the design patterns and principles we discussed when we covered networking—it's just another way to implement them.

For example, let's try to implement the delta updates design pattern using the Combine framework:

```
URLSession.shared.dataTaskPublisher(for: request)
    .tryMap { output in
        guard let response = output.response as?
          HTTPURLResponse, response.statusCode ==
          200 else {
            throw URLError(.badServerResponse)
        }
        return output.data
    }
    .decode(type: ContactsDeltaUpdateResponse.self,
      decoder: JSONDecoder())
    .receive(on: DispatchQueue.main)
    .sink(receiveCompletion: { completion in
        switch completion {
```

```
            case .finished:
                break
            case .failure(let error):
                print("Error during sync:
                    \(error.localizedDescription)")
            }
        }, receiveValue: { [weak self] response in
            self?.processDeltaUpdates(response:
              response)
        })
        .store(in: &cancellables)
```

Looking at the code example, we can see that it looks pretty much like the previous Combine code—that's part of the idea of consistent interface and modular code. We perform the request, check the response code, decode it, change it to the main thread, and process the response data.

Summary

Connecting to our backend and retrieving data is a basic task in most mobile apps. Doing so lets us present valuable and interesting information to our users.

Performing a simple request is easy – however, there are many other factors to bear in mind, and doing that properly is crucial to having an efficient app.

This chapter reviewed the different network components, such as the request itself, error handling, and data storage. We also discussed our different design patterns to work with our backend. We ended up incorporating Combine into our flows. We should now be perfectly able to set up a fantastic network infrastructure for our app.

Now, let's flip to the other side of our architecture, the UI, and discuss a library that can enrich our app easily – **Charts**!

9
Creating Dynamic Graphs with Swift Charts

Swift Charts is a framework by Apple that allows us to present data in beautiful and expressive charts. Working with charts is not a minor topic – data is an essential topic in mobile apps, and the ability to show glance information of insights and trends is crucial to our app's user experience.

In this chapter, we will cover the following topics:

- Understanding why we need charts in our apps
- Meeting the Swift Charts framework
- Creating charts such as bar, line, pie, area, and point charts
- Visualizing functions with Charts
- Implementing user interaction to our charts using ChartProxy
- Allowing different data types to work with charts by conforming to the Plottable protocol

Before we create our first chart, let's understand why charts are important and what value they bring.

Technical requirements

For this chapter, you must download Xcode version 15.0 or above from Apple's App Store.

You'll also need to run the latest version of macOS (Ventura or above). Simply search for Xcode in the App Store and select and download the latest version. Launch Xcode and follow any additional installation instructions that your system may prompt you with. Once Xcode has fully launched, you're ready to go.

Download the sample code from the following GitHub link: `https://github.com/PacktPublishing/Mastering-iOS-18-Development/tree/main/Chapter9/Chapter9.swiftpm`

Why charts?

The following is not necessarily a mobile-specific section but an important one nevertheless. Many apps display helpful information in a textual way, such as tables, lists, or grids. While displaying information in a list or a grid can be beneficial, it's much harder to tell the story that way.

Users sometimes struggle to process a textual representation of information, and visualizing it may help them gain interesting insights and make decisions. There might be different types of insights, which can be relationships between data points, trends, and repeated patterns.

Data can be even more difficult to digest on a mobile phone due to the screen size and the challenge of presenting information in grids. However, screen size is not the only challenge with mobile phones – users often expect to glimpse data insights rather than analyze spreadsheets. A mobile user experience differs from a desktop one because of different use cases and behavior. Due to that difference, charts have even greater value on mobile than on desktop apps, as they provide a way to present information visually.

Having said that, it is essential not to overuse charts or to use charts where a table or a list makes more sense. For example, a banking app that shows the user's latest transactions would use a list rather than a chart. A list is a great way to present raw data in a scannable format that is also interactive and allows users to perform actions or view more details.

Just as we have Lists, Tables, and Collection Views, we now have Swift Charts, a framework dedicated to presenting data in an informative, visualized way.

Introducing the Swift Charts framework

Creating charts that are simple and easy to use was always a challenge. Unlike Tables, Collection views, or Lists, most third-party chart frameworks never felt natural in **UIKit/SwiftUI**.

In iOS 16, Apple announced Swift Charts, a SwiftUI framework that presents structured data in a chart and fits nicely in a SwiftUI view.

Let's see an example of a bar chart:

```
import Charts
struct BarMarkView: View {
    struct Sales: Identifiable {
        var id: UUID = UUID()
        let itemType: String
        let qty: Int
    }

    let data: [Sales] = [
        Sales(itemType: "Apples", qty: 50),
```

```
            Sales(itemType: "Oranges", qty: 60),
            Sales(itemType: "Watermelons", qty: 30)
    ]

    var body: some View {
        VStack {
            Chart(data) {
                BarMark(
                    x: .value("Fruit", $0.itemType),
                    y: .value("qty", $0.qty)
                )
            }
        }
    }
}
```

Even though the code example seems long, it is simple to read and understand. This example displays a BarMark chart showing different sales figures for fruits. It has a Sales structure that contains a single sales information for a specific fruit type and a data array that contains sales information about several fruit types.

In the SwiftUI body part, we add a new view called Chart with the data array as a parameter. Inside that Chart view, we add a BarMark view – a way to present data information in bars – passing the x and y values from our Sales struct.

Figure 9.1 shows the result:

Figure 9.1: A BarMark chart

Figure 9.1 shows our code result—a view with three red bars, including a legend and titles. We can see how much easier it is to create a chart, similar to how we would make a `List` or a `VStack` view.

Let's explore and learn how to create the different chart types and understand their usage.

Creating charts

Before we continue, let's understand the view structure of a chart in the Swift Charts framework. As we can see from the last code example, the chart view is called `Chart`:

```
Chart(data) {
        BarMark(
            x: .value("Fruit", $0.itemType),
            y: .value("qty", $0.qty)
        )
    }
```

Each data point in the chart is called a **Mark**. In this code example, we have a Chart with three data points (marks) of the `BarMark` type. If the Chart receives an array as a parameter, it performs a `ForEach` loop under the hood and creates several marks.

In fact, we could write the same code as the following:

```
Chart {
    ForEach(data, id:\.id) { item in
        BarMark(x: .value("Fruit",
                    item.itemType),
                y: .value("qty", item.qty))
    }
}
```

In this code example, we take the same data array as before, iterate it using a `ForEach` loop, and create a `BarMark` view for each array item. This example is crucial to understanding how charts are built so we can customize and configure them in the future.

Now, let's explore the `BarMark` chart even further.

Creating BarMark chart

We can use a `BarMark`-based chart to compare different data points, such as sales figures and country population sizes. We saw how simple creating a chart with multiple bar marks is.

However, implementing a chart with BarMark views doesn't end here. We have more options to expand that mark to provide even more information.

We'll start with a stacked bar chart.

Adding Stacked Marks

Standard marks represent two-dimensional data points, comparing one value to another. Sometimes, datasets may have a deeper story, as each bar may be constructed from several values.

For example, let's take the sales chart we have just created and discuss the sales of apples. The current value of apple sales is 50 items. Perhaps we want to display how this value is divided between green and red apples. In this case, we can use a stacked mark.

We will now add a stacked bar to our existing chart.

First, we need to adjust our `Sales` structure to contain our fruit color:

```
struct Sales: Identifiable {
    var id: UUID = UUID()
    let itemType: String
    let qty: Int
    var fruitColor: String = ""
}
```

Now that we have added a `fruitColor` property to the `Sales` structure, we can update our dataset:

```
let data: [Sales] = [
        Sales(itemType: "Apples", qty: 20, fruitColor:
          "Green"),
        Sales(itemType: "Apples", qty: 30, fruitColor:
          "Red"),
        Sales(itemType: "Oranges", qty: 60),
        Sales(itemType: "Watermelons", qty: 30)
]
```

Currently, our updated dataset has two records related to apple sales, each containing the color sold.

Now that we have all the data that we need, let's create a chart and assign each of the properties to the right role in the chart:

```
Chart(data) {
        BarMark(x: .value("Fruit", $0.itemType),
                y: .value("qty", $0.qty))
          .foregroundStyle(by: .value("Color",
            $0.fruitColor))
    }
```

In this code example, the only difference we have is the `foregroundStyle` view modifier, which helps distinguish between the different fruit colors. Let's see the result in *Figure 9.2*:

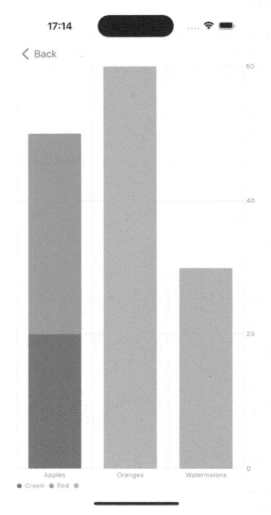

Figure 9.2: Stacked bar view

In *Figure 9.2*, we can see that the apples bar is built from two types of values. The blue represents green apples, and the green represents red apples.

We saw that when we add several marks with the same *x* values, the Charts framework knows how to stack them together.

Next, let's see what happens when we don't add *y* values to our data set.

Adding 1D bar marks

Most charts are two-dimensional, meaning they have an *x* and *y* axis that compares different data categories. However, we can focus on one category (meaning the chart will have only one *y* axis value) and create a one-dimensional chart.

For instance, let's take the apple category from the previous example and try to create a 1D bar based on it.

First, let's enrich our data and add `Yellow` as an additional fruit color:

```
let data: [Sales] = [
    Sales(itemType: "Apples", qty: 20, fruitColor:
        "Green"),
    Sales(itemType: "Apples", qty: 30, fruitColor:
        "Red"),
    Sales(itemType: "Apples", qty: 40, fruitColor:
        "Yellow"),
]
```

Our dataset now includes the `Green`, `Red`, and `Yellow` fruit colors.

Next, let's create our chart, but this time, we won't define the *y*-axis:

```
Chart(data) {
            BarMark(
                x: .value("Qty", $0.qty)
            )
            .foregroundStyle(by: .value("Color",
                $0.fruitColor))
        }
```

In this code example, we passed only the x `BarMark` parameter. However, if we examine the `BarMark` header, we can see that there's a method that requires only the x parameter:

```
public init<X>(x: PlottableValue<X>, yStart: CGFloat? = nil,
yEnd: CGFloat? = nil, width: MarkDimension = .automatic, stacking:
MarkStackingMethod = .standard) where X : Plottable
```

The `init()` function in this code example is the method that we are using. Now, let's see what the chart we create looks like when it's only one-dimensional (*Figure 9.3*):

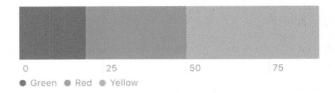

Figure 9.3: A 1D chart

In *Figure 9.3*, our data is presented in a one-dimensional chart presenting three different types of apples.

One thing still bothers us here: notice that the fruit colors don't match the actual colors the Charts framework assigned to each fruit when it created the chart. That's because the Charts framework generates the colors while encoding the value. If we want to match the fruit color to the chart presented color, we need to use the `chartForegroundStyleScale` view modifier:

```
Chart(data) {
            BarMark(
                x: .value("Qty", $0.qty)
            )
            .foregroundStyle(by: .value("Color",
              $0.fruitColor))
        }
        .chartForegroundStyleScale(["Green" :
          Color.green, "Red" : Color.red,
          "Yellow" : Color.yellow])
```

The `chartForegroundStyleScale` function is a view modifier we can apply to the Chart and different `ShapeStyle` protocol to different values. In this case, we use colors that reflect the fruit colors and improve clarity.

Figure 9.4 shows how the chart looks now that we matched the colors to the names:

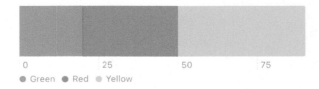

Figure 9.4: A 1D chart with custom colors

We can use `chartForegroundStyleScale` not only for 1D charts but also for all other types of charts.

We saw how to use BarMarks for stacked and one-dimensional marks. Yet another way we can use BarMarks is for interval bar charts.

Adding interval bar charts

We use **interval bar charts** to represent data grouped into intervals, such as periods, age groups, or numerical ranges.

For example, let's say we want to display a list of workers and the time intervals they worked throughout the day.

First, let's create a data set that represents a list of working periods:

```
let emma = "Emma Johnson"
let liam = "Liam Patel"
let sophia = "Sophia Garcia"

let data: [EmployeDayWork] = [
        EmployeDayWork(name:emma, startTime: 10, endTime:
          12),
        EmployeDayWork(name:liam, startTime: 8, endTime:
          11),
        EmployeDayWork(name: sophia, startTime: 10.5,
          endTime: 11.5),
        EmployeDayWork(name: emma, startTime: 14, endTime:
          15),
        EmployeDayWork(name: liam, startTime: 13.5,
          endTime: 14.2),
        EmployeDayWork(name: sophia, startTime: 15,
          endTime: 16)
]
```

Each item in the `data` array represents one employee's working period. Notice that we don't care about the item's order—the Charts framework is responsible for ordering them correctly. However, we care about consistency with the employee's name, so the Charts framework can also properly group the items.

Let's see how we can build an interval chart based on that dataset:

```
Chart(data) {
            BarMark(
                xStart: .value("Start", $0.startTime),
                xEnd: .value("End", $0.endTime),
```

```
                    y:    .value("Employee", $0.name)
              )
          }
```

In this code example, we create a BarMark initializer that includes new parameters—`xStart`, which represents the value where the interval begins, `xEnd`, detailing where it ends, and `y`, the employee's name.

Now, let's see how an interval chart looks when we run it (*Figure 9.5*):

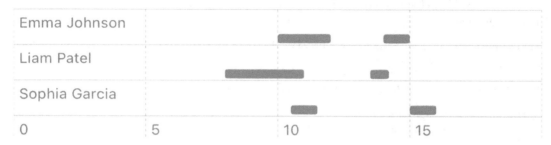

Figure 9.5: An interval chart

In *Figure 9.5*, we can see a timeline when each of the employees is represented in a row, and their working periods are intervals in this timeline. The interval bar chart is an excellent example of a component that can be complex to build from the ground up, and the Charts framework can simplify the process.

BarMark seems like a very flexible chart type, and that's part of the reason it is so common. It allows us to present different information types, whether comparing values or different trends over time, in stacked, one-dimensional, or interval layouts.

However, sometimes, it's a better choice to pick a more specific chart that expresses data more precisely.

So, let's meet the LineMark chart.

Creating LineMark charts

One of the challenges of presenting data in a table is showing trends and patterns. Even though the BarMark chart type can do that better than a table, there are better ways to show trends, especially when dealing with a large amount of information.

To show trends and patterns more efficiently, we can use the LineMark chart, which represents data using a line representing a list of data points.

Let's take, for example, a chart that shows phone sales over time. We create a structure named `SalesFigure` that contains information about the product type, the day of the sales, and the total amount:

```
struct SalesFigure: Identifiable {
    var id: UUID = UUID()
    let product: String
    let day: Date
    let amount: Double
}
```

Now that we have a structure, let's create our dataset like we did in all previous examples:

```
let phoneProduct = "Phone"

let salesFigures: [SalesFigure] = [
        SalesFigure(product: phoneProduct, day:
          Date(timeIntervalSince1970: 1714078800), amount:
            100),
        SalesFigure(product: phoneProduct, day:
          Date(timeIntervalSince1970: 1714165200), amount:
            120),
        SalesFigure(product: phoneProduct, day:
          Date(timeIntervalSince1970: 1714251600), amount:
            90),
        SalesFigure(product: phoneProduct, day:
          Date(timeIntervalSince1970: 1714338000), amount:
            70)
    ]
```

The `salesFigures` variable contains information about four days of sales. The LineMark chart is suitable for working with many entries, but we use only four for demonstration purposes.

Now, let's connect the `salesFigures` variable to a chart using the `LinkMark` view:

```
Chart(salesFigures) {
            LineMark(
                x: .value("time", $0.day),
                y: .value("amount", $0.amount)
            )
        }
```

We created a LineMark inside the chart, setting the day as the *x* axis and the amount as the *y* axis. Running that code should show us a chart that looks like *Figure 9.6*:

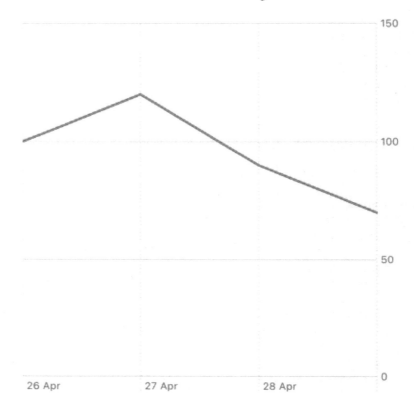

Figure 9.6: A LineMark chart

The chart in *Figure 9.6* shows the declining trend of phone sales over the dataset period. What's nice about line charts is that it's easy to compare one LineMark to another. All we need to do is to update our dataset. So, let's also add tablet sales to compare it with phone sales:

```
let tabletProduct = "Tablet"
let salesFigures: [SalesFigure] = [
        SalesFigure(product: phoneProduct, day:
          Date(timeIntervalSince1970: 1714078800), amount:
            100),
        SalesFigure(product: tabletProduct, day:
          Date(timeIntervalSince1970: 1714078800), amount:
            70),
    // …
```

```
        SalesFigure(product: phoneProduct, day:
          Date(timeIntervalSince1970: 1714338000), amount:
            70),
        SalesFigure(product: tabletProduct, day:
          Date(timeIntervalSince1970: 1714338000), amount:
            110)
    ]
```

In this code example, we updated our dataset by adding tablet sales figures items to the array.

To make the chart distinct between the two product types, we use the `foregroundStyle` view modifier:

```
LineMark(
     x: .value("time", $0.day),
     y: .value("amount", $0.amount)
     ).foregroundStyle(by: .value("Product", $0.product))
```

Adding the `foregroundStyle` view modifier applies different styles to different product types. Looking at the code, we can see that the chart can distinguish between these two types.

Let's see what the chart looks like after we have added the tablet sales figures (*Figure 9.7*):

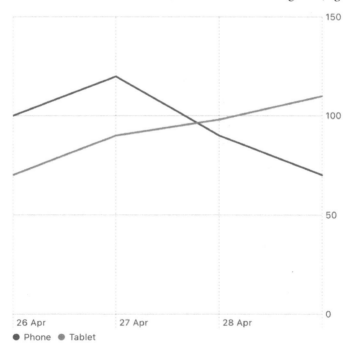

Figure 9.7: LineMark chart with two types of product sales figures

Figure 9.7 shows tablet sales compared to phone sales. We can see that while the phone sales declined, the tablet sales increased. That's an insight that is difficult to get just from the dataset.

Thus far, we have created two primary types of charts: bar and line charts. These two types are pretty popular, as they are simple to understand and work for many use cases.

Another popular chart type Apple added in iOS 17 is **SectorMark**, also known as a pie chart.

Creating a SectorMark chart

A SectorMark, or pie, chart provides a way to visualize the proportions of different values. Unlike the other charts, the pie chart is based on a circular shape divided into slices, and each slide represents a different item value.

Apparently, the SectorMark chart looks like another form of Stacked Marks we covered earlier (under *Adding Stacked Marks*).

However, SectorMark charts became more popular than Stacked Marks as they are visually appealing and easier to understand. Moreover, StackedMark and SectorMark charts are excellent for comparing different parts and seeing their contribution to the whole. However, stacked marks are practical when we want to compare one whole to another, and SectorMark charts are helpful when we want to focus on one whole.

Like the previous examples, to create a SectorMark chart, we need to prepare a dataset. So, let's create a dataset representing a poll result about consuming fruits:

```
let data: [FavoriteFruit] = [
    FavoriteFruit(name: "Apple", value: 30),
    FavoriteFruit(name: "Banana", value: 25),
    FavoriteFruit(name: "Orange", value: 20),
    FavoriteFruit(name: "Strawberries", value: 15),
    FavoriteFruit(name: "Grapes", value: 10)
]
```

In this example, we created a structure named `FavoriteFruit`, which contains the name of the fruit and the number of people who chose that fruit.

To use the data dataset, we will add a `SectorMark` view to our chart:

```
Chart(data) {item in
    SectorMark(angle: .value("Value", item.value))
        .foregroundStyle(by: .value("Fruit",
        item.name))
}
```

The `SectorMark` structure has an angle parameter that reflects the numeric value of the slice. We also added the `foregroundStyle` view modifier, which colors the slice according to the item's fruit property.

Let's look at what the SectorMark chart looks like when running our code (*Figure 9.8*):

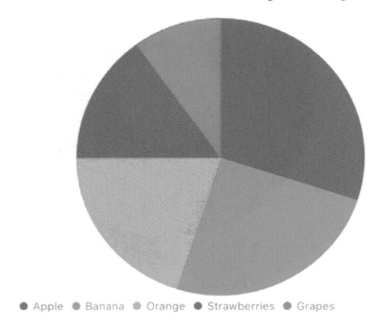

Figure 9.8: SectorMark chart

Figure 9.8 shows a beautiful, colorful pie chart, including the legend titles. We can even set an inner radius to add a **donut style** to the chart:

```
Chart(data) {item in
    SectorMark(angle:  .value("Value", item.value),
        innerRadius: 50)
        .foregroundStyle(by:  .value("Fruit", item.name))
}
```

The addition of the inner radius creates a **hole** in the pie chart, as we can see in *Figure 9.9*:

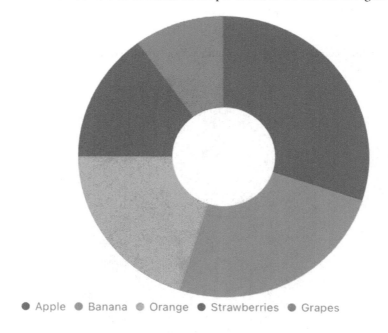

Figure 9.9: A SectorMark chart with an inner radius

Figure 9.9 shows a donut-shaped SectorMark chart. This shape allows us to provide more information in the center of the chart. Some even say that this form is more readable to users as it eliminates the need to compare angles.

Until now, we have created `BarMark`, `LineMark`, and `SectorMark` charts. The following chart combines two charts we discussed – the LineMark and stacked BarMark charts. That's the `AreaMark` chart.

Creating an AreaMark chart

The stacked BarMark chart we discussed under the *Adding Stacked Marks* section shows two important figures – the total value of a category and how that total is divided into sub-categories while observing the different proportions. The LineMark chart, on the other hand, shows the trend or patterns between different data points.

However, what if we want to combine these two types of marks, showing how a value is composed of different categories over time?

That's what the AreaMark chart is for.

Let's take our LineMark sales figures example. We have a dataset representing phone and tablet sales over time. Now, we want to see the total sales of these two types of products over time while still observing the different trends of each product.

So, we can create an `AreaMark` chart based on the same dataset:

```
Chart(salesFigures) { data in
        AreaMark(
            x: .value("Date", data.day),
            y: .value("Sales", data.amount)
        )
        .foregroundStyle(by: .value("Product",
          data.product))
    }
```

Our code example is identical to the LineMark example we discussed under the *Creating LineMark charts* section; the only difference is that we are now using AreaMark instead of LineMark.

However, the result is different (*Figure 9.10*):

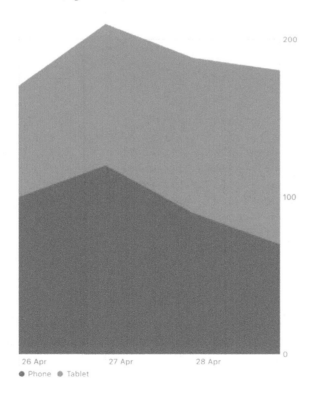

Figure 9.10: An AreaMark chart for total sales

At first glance, *Figure 9.10* shows similar information as *Figure 9.7*—trends of product sales figures. However, there are differences. The filled area in *Figure 9.10* represents the **total sales** of products for both phones and tablets, and each color represents a different product type. On the other hand, *Figure 9.7* only shows a comparison between these two product types, side by side.

The AreaMark chart is great for market share analysis, financial data visualization, and general information, including data trends and cumulative totals.

However, charts can give us much more than data comparison and trends. Let's meet our final chart, PointMark, which can provide a different level of insight.

Creating a PointMark chart

Until now, we have discussed marks that have helped us compare sales figures or observe trends. What about areas such as correlation analysis or predictive modeling? To fulfill that need, the PointMark chart, also known as the **scatterplot chart**, aims to show the relationships between two variables.

Let's find the correlation between students' study hours and grades. First, we create a dataset representing the data:

```
struct StudentData: Identifiable {
    var id: UUID = UUID()
    var hoursStudied: Double
    var examScore: Double
}

let studentDataSet: [StudentData] = [
    StudentData(hoursStudied: 1.7, examScore: 61.8),
    StudentData(hoursStudied: 7.9, examScore: 78.6),
    StudentData(hoursStudied: 4.1, examScore: 44.3),
    StudentData(hoursStudied: 4.7, examScore: 63.4),
    StudentData(hoursStudied: 7.8, examScore: 90.4),
    StudentData(hoursStudied: 8.6, examScore: 83.2),
    StudentData(hoursStudied: 2.8, examScore: 29.7),
    StudentData(hoursStudied: 6.3, examScore: 72.9),
    StudentData(hoursStudied: 6.4, examScore: 73.8),
    StudentData(hoursStudied: 6.1, examScore: 77.6)
]
```

This code example has a `StudentData` structure containing information about student study time and grades. `studentsDataSet` is an array that contains information about ten students.

Now, let's create a `PointMark` chart based on that array:

```
Chart(studentDataSet) {
    PointMark(x: .value("hours", $0.hoursStudied),
```

```
                y: .value("score", $0.examScore))
}
```

Like previous charts, the `PointMark` structure has `x` and `y` parameters. The `x` parameter represents the hours studied, and the `y` parameter represents the score.

Figure 9.11 shows what the `PointMark` chart looks like when running the code:

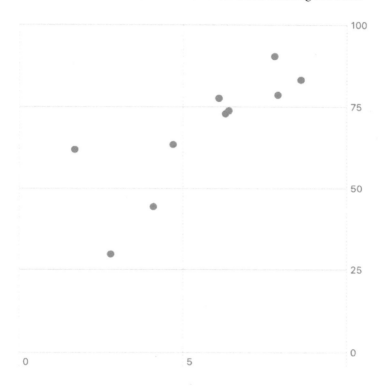

Figure 9.11: PointMark chart

Figure 9.11 shows that most students achieve high grades when studying more hours. We can also identify one student who managed to achieve a mid-level grade almost without studying at all!

Even though PointMark is less common than the previous charts we reviewed, it can be helpful in financial, CRM, or education apps.

Speaking of education apps, many apps require other types of charts. That includes charts that are based on functions and not datasets. With Charts, we can also work more dynamically and straightforwardly visualize functions. Let's see how to do that.

Visualizing functions with Charts

Until now, we have discussed how to build charts using datasets, which include raw data information such as sales figures, market shares, or usage trends. However, we don't have to use datasets to create charts, as functions can also perform as a data source for our charts.

For example, we may want to display a normal distribution line graph next to our BarMark chart. We could also create an education app that displays mathematical functions such as circles or a sinus function.

To do that, we need to use a different type of chart called **plot**.

The Charts framework has two types of plots – `LinePlot` and `AreaPlot`. Let's see an example of `LinePlot` showing a graph for a sinus function:

```
Chart {
    LinePlot(x:"x", y:"y") { x in
        return sin(x)
    }
}
```

In this (very!) short code example, we added a `LinePlot` chart with a closure that returns the `y` value of a given `x` value. In this case, we used a simple `sin` function. *Figure 9.12* shows what the chart looks like:

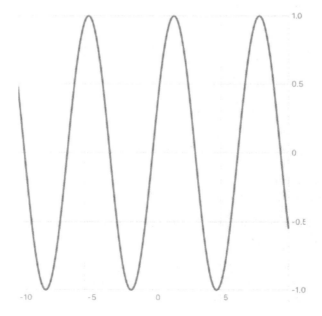

Figure 9.12: A LinePlot chart

In *Figure 9.12*, we can see the `LinePlot` chart generated from a simple mathematical function.

As mentioned earlier in this section, the second chart type we can use to visualize functions is `AreaPlot`, the equivalent of `AreaMark`:

```
Chart {
    AreaPlot(x:"x", y:"y") { x in
        return sin(x)
    }
}
```

In this code example, we only changed the chart type from `LinePlot` to `AreaPlot`. `AreaPlot` visualizes the function by filling the area it defines. Let's see the output in *Figure 9.13*:

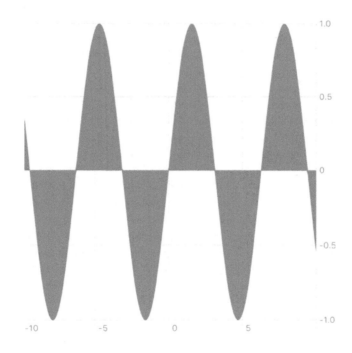

Figure 9.13: The AreaPlot chart type

Figure 9.13 shows the same sinus function graph, now filled with color.

Using the LinePlot and AreaPlot chart types to visualize math functions is about much more than just showing how the sinus function behaves. It is excellent for education, scientific research, finance, health, and business apps. Now that we know how to create LinePlot and AreaPlot, we have whole new options.

We went over many chart types, and by now, we can quickly create charts, just like creating a simple list!

The **List** type provides a way to interact with its items, allowing us to navigate or delve into more information. So, let's see how to make our charts interactive!

Allowing interaction using ChartProxy

Now that we know how to create charts, let's discover more hidden tricks by adding user interaction capabilities. User interaction in charts, with its many uses, allows users to explore the chart's data using touch. Here are some use cases for user interaction with charts:

- **Drill down to a specific data mark**: By touching a `BarMark` or `SectorMark` charts, the user can navigate to a new screen that shows additional information about the particular data point. For example, if the `BarMark` chart shows information about watermelon sales, we can navigate the user to a screen that details the watermelon sales deals.
- **Exploring new data points**: Enabling user interaction with `LineMark` charts, for example, provides insights to the user on data points not originally part of the dataset if our `LinkMark` chart includes information about the growing population in a specific city over time, touching a particular point the chart can display the population value on a specific date.
- **Comparing data marks**: The user can highlight and compare multiple data marks, which is extremely useful in BarMark-based charts.

Moreover, learning how to add interaction capabilities can help us explore more things with our charts, such as how the charts are built and how their calculation logic works.

To understand how interaction works, we need to get to know more Swift Charts framework components:

- `chartOverlay`: This is a view modifier that helps us add an overlay view to a chart. We can use the `chartOverlay` view modifier to add more graphic details to our chart, such as rulers and texts. We can also use the `chartOverlay` view modifier to observe gestures and user interaction.
- `ChartProxy`: This is a proxy that lets us access the chart values based on the chart area. Using `ChartProxy`, we can convert locations to values and vice versa.

`ChartOverlay` and `ChartProxy` are essential components when handling user interaction; therefore, they come hand in hand. When adding a `chartOverlay` view modifier, it comes with a proxy to have complete access to the chart.

Let's try to take a LineMark chart and add a horizontal ruler that allows users to drag their fingers across it. We'll start by adding an overlay.

Adding an overlay to our chart

The solution for providing an overlay to our chart consists of a common practice in SwiftUI using a view modifier. Look at the following code example:

```
Chart(salesFigures) {
        LineMark(
            x: .value("time", $0.day),
            y: .value("amount", $0.amount)
        )
        .foregroundStyle(by: .value("Product",
           $0.product))
    }
    .chartOverlay { proxy in
    }
```

We took the LineMark example from the *Creating LineMark charts* section and added a `chartOverlay` view modifier in this code example.

We can see that `chartOverlay` comes with a `proxy` variable, which is the `ChartProxy` component we discussed earlier.

`ChartOverlay` is not a view but a view modifier that lets us add new views to the chart. So, to recognize gestures and add a ruler, we can add a transparent view with a drag gesture and add a ruler view:

```
.chartOverlay { proxy in
    ZStack(alignment: .topLeading) {
        Rectangle().fill(.clear)
            .contentShape(Rectangle())
            .gesture(
                DragGesture()
                    .onChanged { value in
                    }
            )
        let lineHeight = proxy.plotSize.height
        Rectangle()
            .fill(.red)
            .frame(width: 2, height:
              lineHeight)
            .position(x: markerX, y:
              lineHeight/2)
    }
}
```

In this code example, we added a `ZStack` view with a clear rectangle that covers the whole chart and, on top of it, a red ruler view. The ruler view *x* axis is a state variable:

```
@State var markerX: CGFloat = 50
```

We are going to change it according to the user's tap locations.

Notice that we used our `proxy` object to determine the chart size for the ruler view. This is crucial proxy usage, as we will need it on other occasions, such as calculations for displaying different views in particular locations.

To see our view structure, look at *Figure 9.14*:

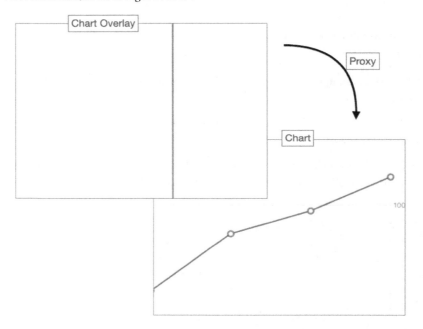

Figure 9.14: Chart and chartOverlay structures

Figure 9.14 shows our chart view and the rectangle we added using the `chartOverlay` view modifier. We can also see that they are connected using the proxy object.

Also, we added a drag gesture to the rectangle. Let's see how to use it to change our ruler position accordingly.

Responding to the user's gesture

To respond to the user's gesture and move the horizontal ruler to the closest data point, we need to implement the onChanged closure:

```
.onChanged { value in
   markerX = value.location.x
   if let closestDate = getClosestDateForLocation(x:
     value.location.x, proxy: proxy) {
       if let positionX = proxy.position(forX:
         closestDate) {
            markerX = positionX
     }
   }
}
```

The onChanged closure implementation does three things:

- First, it *finds the closest sales data point* according to the tap location and the proxy. We will go over the getClosestDateForLocation function in a minute.
- After we found the closest sales data point according to the tap location, we used the proxy object to *retrieve its position* on the chart. One of the proxy's capabilities is to convert data points to position and vice versa.
- When we have the position of the closest data point, we adjust the ruler location by setting the markerX state variable.

This piece of code is a good demonstration of what we can do with the proxy object.

For more proxy object usage, let's see the getClosestDateForLocation function.

Finding the closest data point to the user's touch

The getClosestDateForLocation function goal is to find the closest date with a data point according to a specific position.

The function receives two parameters – the position (CGFloat) and the proxy object:

```
func getClosestDateForLocation(x: CGFloat, proxy: ChartProxy) -> Date?
{
        var returnedSalesFigure: SalesFigure?
        if let date = proxy.value(atX: x) as Date? {
           var mDistance: TimeInterval = .infinity
           for salesFigure in salesFigures {
              let distance =
```

```
                    abs(salesFigure.day.distance(to: date))
                if distance < mDistance {
                    returnedSalesFigure = salesFigure
                    mDistance = distance
                }
            }
        }
        return returnedSalesFigure?.day
}
```

Remember what our chart looks like – the *y* axis represents the timeline, and the *x* axis represents the total sales on a specific date.

So, we can use the proxy object to find the date for a specific x value, and that's our first step:

```
if let date = proxy.value(atX: x) as Date? {
```

The proxy's `value(atX:)` function calculates the date value for a specific x value.

However, the returned value is arbitrary; to locate the closest data point, we must iterate through our dataset and search for the nearest `SalesFigure` object. Once identified, the function can then return it.

Even though allowing user interaction with charts is not complex, it includes interesting view modifiers and objects that enable us to access the chart data, perform calculations, and display overlay UI components. We don't have to use the `proxy` object and the `chartOverlay` view modifier just for interaction—we can show additional information, improve the chart design, and, in rare cases, even create our chart.

Until now, we used data sets with foundation types – `String`, `Double`, and `Date`. However, when we look at the Swift Charts framework headers, we see something interesting:

```
func position<P>(forX value: P) -> CGFloat? where P :
    Plottable
public struct LineMark {
    init<X, Y>(x: PlottableValue<X>, y: PlottableValue<Y>)
        where X : Plottable, Y : Plottable
}
```

It seems that the different chart functions only work with types that conform to the `Plottable` protocol. Let's find out what that is.

Conforming to the Plottable protocol

Until now, we have been under the assumption that any data set we threw on our charts would work. However, we saw that the proxy object can perform interesting calculations that are not possible with any data, and that's only one reason why our data types need to support the ability to be drawn in a chart.

Therefore, the Swift Charts framework only works with data types that conform to the `Plottable` protocol, which allows data to be drawn in a chart.

First, every primitive data type already conforms to the `Plottable` protocol. Also, the `Date` class that we used in our last example conforms to the `Plottable` protocol. We can even see that in the apple header files:

```
extension Date : Plottable, PrimitivePlottableProtocol

extension String : Plottable, PrimitivePlottableProtocol
```

However, working only with primitive or Foundation types is not always practical.

Let's take, for example, our `Sales` structure from the *Adding Stacked Marks* section:

```
struct Sales: Identifiable {
    var id: UUID = UUID()
    let itemType: String
    let qty: Int
    var fruitColor: String = ""
}
```

Declaring the `itemType` property as a string is not always a best practice. Typically, types are part of a closed list, and using strings may lead to typos and duplicates. We probably would like to use an enum for that, as it is more suitable for handling a list of types:

```
enum FruitType {
    case Apples
    case Oranges
    case Watermelons
}

struct Sales: Identifiable {
    var id: UUID = UUID()
    let itemType: FruitType
    let qty: Int
    var fruitColor: String = ""
}
```

In this example, we created a `FruitType` enum to replace the `itemType` type from `String`.

Our next step is to make the `FruitType` enum conform to `Plottable`:

```
extension FruitType: Plottable {
    var primitivePlottable: String {
        rawValue
    }
}
```

In this example, we used the `primitivePlottable` variable getter to return the type's primitive value. That would make the `FruitType` type eligible to be used inside Charts.

Even though not every type can be used inside a chart, we can easily make them eligible for that. Conforming to the `Plottable` protocol is simple and straightforward and allows us to use our custom-made types within charts.

Summary

The Swift Charts framework is exciting. It allows us to create amazing-looking charts using a simple data set, making it much easier to display data insights, trends, and comparisons.

This chapter reviewed the different chart types of the Swift Charts framework, including BarMark, LineMark, SectorMark, AreaMark, and PointMark.

We also discussed each chart's different usage and goals, learned how to customize them, and added user interaction to add more capabilities. At last, we went over the `Plottable` protocol, which allows our charts to use almost any data type we want. By now, we should be able to implement charts in our apps quickly.

Our next chapter includes an advanced yet very powerful topic – Swift macros.

Part 2: Refine your iOS Development with Advanced Techniques

In this part, you will step up your iOS development and explore advanced topics, such as Swift macros, testing, Combine, architectures, **machine learning** (**ML**), and AI. This part is a must-read if you want to get the most from iOS 18.

This section contains the following chapters:

- *Chapter 10, Swift Macros*
- *Chapter 11, Creating Pipelines with Combine*
- *Chapter 12, Being Smart with Apple Intelligence and ML*
- *Chapter 13, Exposing Your App to Siri with App Intents*
- *Chapter 14, Improving the App Quality with Swift Testing*
- *Chapter 15, Exploring Architectures for iOS*

10
Swift Macros

Developers frequently encounter various challenges with their IDEs, often related to missing functionalities, mostly about missing functionalities. With each new Xcode or Swift version, Apple introduces additional features that enhance productivity and simplify tasks. However, even Apple has a hard time fulfilling our needs and demands. Fortunately, this time, we can create customized functionalities using Swift Macros.

Swift Macros is an exciting new feature added to Xcode 15 and iOS 17, and this chapter will help us increase our productivity by achieving more from our IDE.

In this chapter, we will cover the following topics:

- Learning about Swift Macros
- Exploring the `SwiftSyntax` library, which stands behind Swift Macros
- Creating our first Swift macro
- Handling errors and providing more clarity when something goes wrong
- Testing our macro, making sure it runs as expected over time

But now, let's start with the basics and discover Swift Macros.

Technical requirements

You must download Xcode version 16.0 or above for this chapter from Apple's App Store.

You'll also need to run the latest version of macOS (Ventura or above). Search for Xcode in the App Store and select and download the latest version. Launch Xcode and follow any additional installation instructions that your system may prompt you with. Once Xcode has fully launched, you're ready to go.

What is a Swift macro?

You probably heard the term "macro" before in the context of programming. That's perhaps because programming languages such as C/C++ have macros as well.

A **macro** is a structure that lets us define a code pattern that is being replaced by the compiler with a specific set of instructions.

Let's see a short C example:

```
#define SQUARE(x) ((x) * (x))
int num = 5;
int result = SQUARE(num);
```

In our preceding code, we declare a macro called SQUARE that receives one parameter named X, and our compiler replaces it with (x) * (x).

The initial question that comes to mind is this: why? Can't we just define a function?

So, in this case, a simple function that calculates a number's square can be helpful here.

But a macro's primary goal is not to replace functions, as they are great for several reasons:

- **Code reuse**: Notice that code reuse is not "functionality reuse." Code reuse is where we take an actual code snippet and reuse it in different places. For example, if we constantly repeat the same line sequence when declaring a class, a macro can help us avoid repeating ourselves.
- **Improve abstraction**: Macros can help us add another abstraction layer to our code. Imagine writing a macro that generates functions declaration. That's another level we can construct our code.
- **Performance**: In some cases, macros can help us optimize our code. Sometimes, the trade-off between optimization and readability/simplicity can be solved using a macro. A macro can generate a piece of harder-to-read code and yet be optimized. One feature that macro can optimize code for is **loop unrolling** – a way to iterate a loop faster with instruction-level parallelism. Loop unrolling produces less readable code but is much quicker.

In the bottom line, a macro is just a tool that replaces one code with another and inserts a specific code snippet before the compile time. But C macros are full of issues. They are difficult to test, not type-safety, their errors are not clear enough, and sharing them with other developers is not trivial. As part of Xcode 15, the Swift team released a new tool called **Swift Macros** – the Swift version of macros that lets us create macros more efficiently and elegantly.

Let's go over a simple example of macro usage.

In our project, we want to add a macro that adds a function named `log(issue:String)` to classes and structs. That function prints an issue to our log and adds the class or the struct name. We can call that macro @AddDebugLogger, and we can use it as follows:

```
@AddDebugerLogger
class MyClass {

}
```

In the preceding code, we declared a class named `MyClass` and attached a macro named @AddDebugerLogger, which expands to the following code:

```
class MyClass {
    func printLog(issue: String) {
        #if DEBUG
        print("In class named MyClass - \(issue)")
        #endif
    }
}
```

The macro adds a function named `printLog()`, which prints an issue to the console while mentioning the class name as part of the log message. This serves as an example of primary macro usage, illustrating the capabilities of this tool.

But how is the macro familiar with the class name? How does it generate a new function in the right place inside the class? To answer these questions, we first need to meet `SwiftSyntax`, a library that stands in the heart of Swift Macros.

Exploring SwiftSyntax

`SwiftSyntax` is not a new library, and it's part of Swift's code base from its early beginnings. In fact, Swift Macros is part of `SwiftSyntax`, and it uses its capabilities.

Before we dive into `SwiftSyntax` (and there's enough to dive into it), let's learn about how the Swift compiler works (*Figure 10.1*):

Figure 10.1: The Swift compiler process

Don't fear the different expressions you see in *Figure 10.1*. This figure is a high-level overview of how the compiler takes our source code and generates machine code our device can run (the `*.o` files). We don't have to understand every step in that flow, but knowing how it works is essential, especially where `SwiftSyntax` fits in the process.

Let's go over the steps together:

1. **Parse and abstract syntax tree (AST)**: The compiler takes our source code and builds an AST. The AST represents our code hierarchical structure, including classes, structs, variables, and expressions.
2. **Semantic analysis (sema)**: In this phase, the compiler takes our generated AST and performs semantic analysis. The analysis looks out for semantic issues in our code and goes over issues such as type-checking name resolutions and more (when we see "semantic" issues in our build phase; that's the result of this phase).
3. **Swift Intermediate Language Generation (SILGen)**: In this phase, the compiler generates a representation that captures the semantic structure of the code.
4. **Intermediate Representation Generation (IRGen)**: In IRGen, the compiler takes the SILGen result and converts it to a binary close to machine-level code. This process is done with the help of **Low-Level Virtual Machine (LLVM)**, and the code goes through several optimizations.
5. **LLVM linking**: The LLVM links everything together and prepares our code for the final binary creation.

The process may look scary and complex, but remember that this is a significant enrichment for us as iOS developers and is not required for Swift Macros understanding. I demonstrated it because of the first two steps – parse and AST. Let's talk about them for a second.

Parsing and AST

Parsing Swift code is not an easy task. In addition, building the AST is even more complex.

In the building process, we just saw the parsing, and the AST is handled by the `SwiftSyntax` library. So, when we work with the `SwiftSyntax` library, we have the full compiler capabilities. This means we can parse code, analyze it, and even generate new code like the compiler. The `SwiftSyntax` library is a powerful and essential tool when working with Swift macros because when we think of it, it is what Swift macros are all about – understanding the given code and generating a new one.

We understand that learning SwiftSyntax is a prerequisite for writing Swift macros, so let's dive in.

Setting up SwiftSyntax

`SwiftSyntax` is a **Swift package**, meaning it can be linked easily to an existing iOS or macOS project.

> **What is a Swift package?**
> A Swift package is a unit of code distribution in Swift. It's a way to organize, share, and manage Swift code across different projects.

To play with and learn `SwiftSyntax`, we will create a new project and add `SwiftSyntax` as a Swift package to that project, including a playground. To do so, follow these steps:

1. Let's start with opening Xcode and adding a new project.
2. Then, we'll add our `SwiftSyntax` Swift package by selecting **File | Add Package Dependencies…**.

    Now, we are in the adding dependencies window of Xcode (*Figure 10.2*):

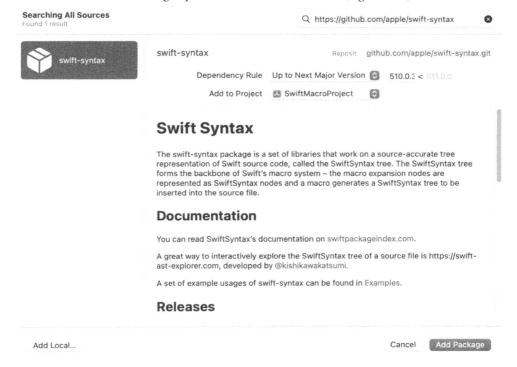

Figure 10.2: The adding dependencies Xcode window

> **Dependencies window?**
>
> If that's the first time you've seen that window, then this is an excellent chance to perform a short introduction. When Swift Package Manager had just started, its management was completely manual, using the Terminal.
>
> Over the years, Swift Package Manager has become an integral part of Xcode, and now, it is even possible to manage collections and search for packages right from Xcode.
>
> You can learn more at `https://www.swift.org/documentation/package-manager/`.

3. Back to Xcode – in the top-right corner of the adding dependencies window, we can fill the `SwiftSyntax` GitHub repository:

   ```
   https://github.com/apple/swift-syntax
   ```

4. We will choose the `swift-syntax` package from the left column and click the **Add Package** button.

5. Xcode will now resolve the Swift package and present its libraries so we can choose what we want to import to our project (*Figure 10.3*):

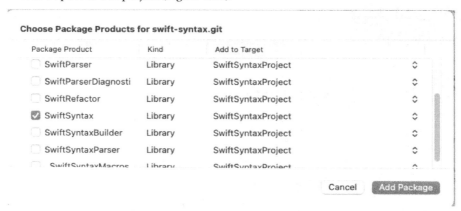

Figure 10.3: Choosing the SwiftSyntax package products

We will choose the `SwiftSyntax` library and click on the **Add Package** button. That's it. We added `SwiftSyntax` to our project!

Now, let's add a playground file (anywhere we like) and explore what `SwiftSyntax` is.

Building our Abstract Syntax Tree

To try and analyze a piece of Swift code using the `SwiftSyntax` library, we need to generate some Swift code and work on it.

We open the playground file we created in the previous section, and add the following code:

```
import SwiftSyntax
import SwiftSyntaxParser

let sourceCode = """
func hello() {
    print("Hello World")
}
"""
```

Our code starts with importing two important libraries – `SwiftSyntax` and `SwiftSyntaxParser`. The `SwiftSyntaxParser` library contains the `SwiftParser` class, which helps convert a source code to a tree we can traverse and analyze.

We added a string constant named `sourceCode` with a simple *"Hello World"* function to see how it works. Imagine that `sourceCode` represents the content of a Swift file.

To parse the "Hello World" code, we'll use `SwiftParser`:

```
do {
    let syntax = try SyntaxParser.parse(source: sourceCode)
} catch {
    print("Error parsing code: \(error)")
}
```

The parsing code is straightforward. `SyntaxParser` calls the parse method with our `sourceCode` constant from earlier and returns a syntax. But what is this syntax? Well, that's our full code tree! The syntax variable is from the type `SourceFileSyntex`, and that type represents the syntax structure of our code. It's the most high-level syntax node, encapsulating all our source code's imports, classes, and functions.

Now, it's time to understand what this syntax tree looks like.

Investigating the tree

One of the best things about working with **Swift Playgrounds** is that it's not only great for playing with code snippets but also for examining their results without having to place breakpoints in our code.

After we run our Playground code, we can see the type `SourceFileSyntax` in the window's right column. When we tap the small square next to it, we can see how the syntax constant is built (see *Figure 10.4*):

Figure 10.4: The syntax object structure

It's an excellent time to take a moment, run it for yourself, and try to understand what we see in *Figure 10.4*. Notice that I marked all the juicy parts.

The syntax instance contains a list of statements. A **statement** is everything we can work with – an import, a class declaration, or even an expression. A statement can contain its own statements.

The base statement class is `CodeBlockItemListSyntax`, and each statement type comes with a different subclass of `CodeBlockItemListSyntax`.

In our case, we have one statement from the type of `FunctionDeclSyntax`, which indicates a function declaration.

Expanding `FunctionDeclSyntax` reveals additional information about the function. For example, its name is represented by the `identifier` property (highlighted with a box in *Figure 10.4*).

`FunctionDeclSyntax` has a `body` property, which contains the property of a statement with all the function statements, including the call for the `print` function.

So, we can see that `SwiftParser` has done all the dirty work for us! Now that we have a tree, we can traverse it. Let's extract the function statement:

```
if let funcDecl = syntax.statements.first?.item.as(FunctionDeclSyntax.
self) {
      // We'll fill that part soon
  }
```

In the preceding code, we are taking the first statement item and trying to convert it to a function declaration type.

There are various declaration types, each providing specific tools to help us traverse and extract more information. Here are some of the most common types we can try to extract:

- `VariableDeclSyntax`: This is for variables
- `EnumDeclSyntax`: This is for enum declaration
- `ClassDeclSyntax`: This is for class declaration
- `ProtocolDeclSyntax`: This is for protocol declaration
- `TypealiasDeclSyntax`: This is for type alias declaration
- `InitialzerDeclSyntax`: This is for construct declaration
- `OperatorDeclSyntax`: This is for operator declaration

These are just some syntax node types available in `SwiftSyntax`, and converting existing statement items to their corresponding types can provide us with the needed functionality.

Let's continue our code example and see what we can get from `FunctionDeclSyntax`:

```
if let funcCallExpression = funcDecl.body?.statements.first?.item.
as(FunctionCallExprSyntax.self) {
   // Checking the print function
   }
```

Let's dissect the preceding code snippet to understand what it accomplishes. With a function declaration, we can dig in and try to analyze the different statements that it contains. In this instance, we can find a statement from the type of `FunctionCallExprSyntax`. This type represents a function call, specifically, a call to `print()`.

Now that we converted the statement to the right type, we can get more information about it:

```
let functionName = funcCallExpression.calledExpression.firstToken?.text
        if functionName == "print" {
            let value = funcCallExpression.argumentList.first?.expression.as(StringLiteralExprSyntax.self)?
                .segments
                .first?.firstToken?.text
        }
```

`funcCallExpression` has a `calledExpression` property that encapsulates the information about actual expression components.

`firstToken` contains the function name itself. But what does "token" mean? Well, **tokens** represent small lexical units of the actual code, such as keywords, variable names, punctuation, or literals. Here, the first token `text` property returns the function name.

Next, we check if the function name is indeed `print`, and now we can check the value being printed by examining the function arguments list. Once we convert the first expression to `StringLiteralExprSyntax`, we can extract its first segment token and store it in the `value` constant.

Does it sound confusing and a little bit cumbersome? Well, we should remember that the `SwiftSyntax` library is not considered easy to work with. It has a steep learning curve with many options and features.

But this complexity is not a coincidence – parsing and analyzing programming language, especially an advanced and full-featured language such as Swift, is not simple. Just like we have `funcCallExpression`, `calledExpression` or `StringLiteralExprSyntax`, we have dozens of different types for different expressions. Looking at the `SwiftSyntax` documentation is the best way to learn to traverse and analyze more of the language.

Now that we understand Swift code analysis using `SwiftSyntax`, let's explore how we can leverage `SwiftSyntax` in the reverse direction – how to generate Swift code.

Generating Swift Code

Generating code in `SwiftSyntax` is based on the built-in types and string literals. We can try and structure Swift code just by creating strings instances:

```
let initString: String = "init(title: String) {
    self.title = title }"
```

We can also generate a piece of Swift code using the `SwiftSyntax` types:

```
let initSyntax = try InitializerDeclSyntax("init(title: String)") {
        ExprSyntax("self.title = title")
    }
```

In the preceding code, `InitializerDeclSyntax` is a constructor declaration, and `ExprSyntax` is a base type for expressions.

In the context of Swift Macros, in most cases, using `String` literals will be enough. That's because the `SwiftSyntax` types support `String` literals. However, using the built-in expressions will ensure the generated code will be valid in future Swift updates.

Speaking of Swift Macros, let's create our first Swift macro now that we know what `SwiftSyntax` is and how it works.

Creating our first Swift macro

As I mentioned earlier (in the *What is a Swift macro?* section), the Swift Macros feature is part of the `SwiftSyntax` library. Macros don't run as part of our app but as a plugin in the IDE.

Macros can be created by adding a new Swift package with a macro template.

It is obvious why Apple selected the Swift package feature to create macros – a Swift package is a great way to encapsulate code, including tests and documentation.

Let's add our first Swift macro by creating a new Swift package.

Adding a new Swift macro

To create a new Swift macro, we should open Xcode and follow these steps:

1. Select **File** | **New** | **Package…**.

2. Then, select **Swift Macro** followed by tapping on **Next** (see *Figure 10.5*):

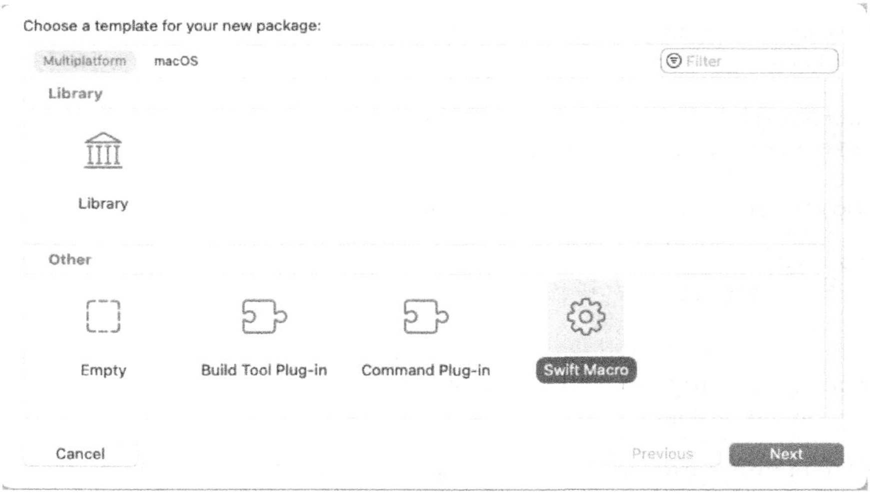

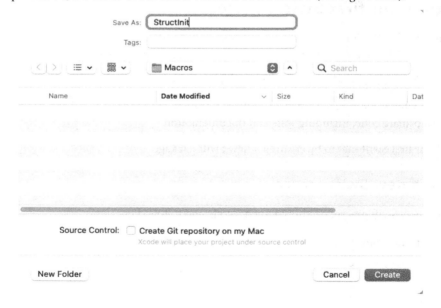

Figure 10.5: Selecting Swift Macro in the choose template window

3. In the opening screen, we will give a name for our struct and press the **Create** button. As part of our learning session, we will create a macro that generates a struct constructor based on its properties. So, the name of our struct will be `StructInit` (see *Figure 10.6*):

Figure 10.6: Adding a StructInit macro

4. After saving, Xcode opens a window with our new package containing an example macro.

Let's see how a Swift Macros package is built next!

Examining our Swift Macros package structure

Now that we have a Swift Macros package, we can reveal its file's structure (*Figure 10.7*):

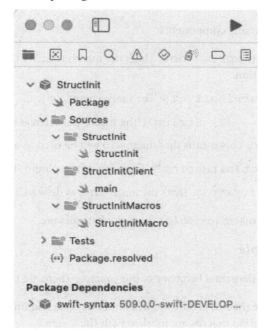

Figure 10.7: The Swift Macros package file's structure

Looking at the Swift Macros package (*Figure 10.7*), we can see that `SwiftSyntax` is defined as a dependency of the package for us, with the latest stable version already linked to our package.

The macro itself is built upon three different source files:

- `StructInit`: That's our macro definition file. Here, we define the macro name and type.
- `StructInitClient`: That's our Swift Macros package executable product. This is where we add an executable code that uses our macro.
- `StructInitMacros`: That's our macro implementation and where all the magic happens.

In addition, we also have a `Test` target where we can test our macro code.

Our first step toward the `StructInit` macro is by declaring its name and type.

Declaring our macro

If we open the `StructInit` file, we can see it has a concise yet important declaration:

```
@freestanding(expression)
public macro stringify<T>(_ value: T) -> (T, String) =
#externalMacro(module: "StructInitMacros",
    type: "StringifyMacro")
```

This short declaration has many components:

- `@freestanding(expression)`: That's the macro role. We'll go over roles in the *Giving our macro a role* section.
- `public macro stringify<T>`: The macro name.
- `(_ value: T) -> (T, String)`: The macro parameters and output.
- `#externalMacro`: This means that the macro will be used as a plug in the compiler.
- `module: "StructInitMacros"`: The name of the plugin that will be used.
- `type: "StringifyMacro"`: That's the macro type, as defined in the `Package.swift` file.

The first component is the macro role, so let's discuss what roles are.

Giving our macro a role

Macro roles define the fundamental behavior of our macros. There are two primary role categories:

- **Freestanding**: These macros can be anywhere in our code and unrelated to a specific class or a function. Freestanding macros are marked with the # sign.

    Here's an example of a freestanding macro:
    ```
    #URL("https://swift.org/")
    ```

    The `#URL` macro checks whether the provided value is a valid URL. If not, it raises an error on compile time. Otherwise, it returns a non-optional value.

    We can see that the `#URL` macro can be anywhere in our code. That's why it is called *freestanding*.

- **Attached**: The second macro type is **attached**. As its name states, the attached macro always comes next to a function, a class, or any other declaration and is marked with a @ sign.

    Here's an example of an attached macro:
    ```
    @StructInit
    struct Book {
        var id: Int
        var title: String
        var subtitle: String
    ```

```
        var description: String
        var author: String
}
```

In the preceding code, the `@StructInit` macro is "attached" to the `Book` struct and inserts an `init` function based on the struct properties.

The two categories of macro types, namely freestanding and attached, represent distinct sets of roles. Here is the list of all roles:

- `#freestanding(expression)`: This just returns a new expression based on an existing one
- `#freestanding(declaration)`: This creates a new declaration
- `@attached(peer)`: This adds new declaration next to the attached one
- `@attached(accessor)`: This adds accessors to a property
- `@attached(memberAttribute)`: This adds attributes to the declarations in the type it's attached to
- `@attached(member)`: This adds new declarations inside the type it's attached to
- `@attached(conformance)`: This adds conformance to the type it's attached to

The role we define when we declare the macro tells the created plugin *how to* change an existing code.

The role is the first part of declaring a macro. Let's continue with the rest of the declaration.

Defining the StructInit macro

Our `StructInit` macro goal is to create the init method for a struct. Our macro doesn't exist independently; its purpose is to insert new declarations into an existing struct. Therefore, we will choose the `@attached(member)` macro from the roles list in the *Giving our macro a role* section:

```
@attached(member)
```

However, mentioning the role type is not enough. We also need to specify what declaration types we expect our macro to generate. In this case, we expect the macro to generate an `init` function. Let's add that to the role declaration:

```
@attached(member, names: named(init))
```

Adding role types helps the compiler cover different cases where the macro generates something else that was not declared. It also behaves as a documentation for our macro.

Here is another example of `names` argument usage:

```
@attached(member, names: named(rawValue))
```

In this case, the `names` argument declares a usage of the `RawValue` declaration.

We can also add `arbitrary` for general purposes:

```
@attached(member, names: arbitrary)
```

Using `arbitrary` counts for all types of declarations.

Moving forward, we will reconfigure the predefined macro with the following declaration:

```
@attached(member, names: named(init))
public macro StructInit() = #externalMacro(module:
    "StructInitMacros", type: "StructInit")
```

Besides adding the role declaration, we also renamed both the macro name and type to `StructInit`.

The macro is short but tells a lot about its goal and behavior. Here comes the important part – the macro implementation.

Implementing the macro

Unlike other Swift types, in macros, we separate our declaration and implementation into different files. In a way, it resembles Objective-C or C++, when the header and the implementation were other parts.

We will open our `StructInitMacros` file and clear its content for a clean start. Afterward, we can proceed to import the relevant libraries:

```
import SwiftCompilerPlugin
import SwiftSyntax
import SwiftSyntaxBuilder
import SwiftSyntaxMacros
```

These are the standard libraries in most macros we will write. Notice that we have `SwiftSyntax` and `SwiftSyntaxBuilder` as part of what we've learned in the *Exploring SwiftSyntax* section.

Now, let's move on to the main dish – the `StructInit` struct.

Declaring the StructInit struct

In Swift Macros, Apple continues its trend of working mainly with structs and protocols instead of classes and inheritance.

To implement a new macro, we will add a new struct with the macro name that conforms to a protocol named MemberMacro:

```
public struct StructInit: MemberMacro {
    public static func expansion() {
        //Implementation details are detailed in the next section
    }
}
```

The compiler looks for a struct with an identical name to the macro name we declared earlier under the *Adding a new Swift macro* section. We also declared the StructInit as public – remember that a macro is part of a Swift package, so we need to have it accessible from other modules as well.

So, what is the MemberMacro protocol? The MemberMacro protocol contains one crucial function that performs the expansion operation, with the non-surprising name of expansion().

However, we won't use MemberMacro every time we create a macro, as it is relevant only to the macro's attached(member) role. Each role has a different protocol we need to conform to.

Here is the list of the different roles and their corresponding protocol:

- @freestanding(expression) -> ExpressionMacro
- @freestanding(declaration) -> DeclarationMacro
- @attached(peer) -> PeerMacro
- @attached(accessor) -> AccessorMacro
- @attached(memberAttribute) -> MemberAttributeMacro
- @attached(member) -> MemberMacro
- @attached(conformance) -> ConformanceMacro

Since we are building a Swift macro with the @attached(member) role, we will focus only on MemberMacro, even though the concept is similar to the other protocols.

Let's go over it together!

Implementing the expansion function

I'll start by showing you the expansion function:

```
public static func expansion(
    of node: AttributeSyntax,
    providingMembersOf declaration: some
      DeclGroupSyntax,
    in context: some MacroExpansionContext
) throws -> [SwiftSyntax.DeclSyntax]
```

While the function may look a little bit complex, we need to remember two things:

1. Most types mentioned in the function signature should already be recognizable to us, as they are components of the `SwiftSyntax` library.
2. There's only one function in this protocol. No need to implement another one!

The `expansion` function aims to receive information about the attached object or the macro parameters and return a piece of Swift code, represented by an array of `SwiftSyntax` expressions (`DeclSyntax`).

The expansion function has three parameters:

1. `node: AtributeSyntax`: This node represents that actual macro in the original piece of Swift code.
2. `declaration: some DeclGroupSyntax`: The declaration struct that describes the struct/class the macro is attached to.
3. `context: some MacroExpansionContext`: The context provides us with more information about the compiler. Remember that the compiler serves as the "environment" in which the macro functions.

Now, we can start creating our struct `init` method.

First, we need to have a list of all the struct properties, including names and types. To do that, we need to analyze the code using `SwiftSyntax`, which we just learned in this chapter (in the *Exploring SwiftSyntax* section).

So, let's get all the struct information that we need:

```
let members = declaration.memberBlock.members // 1
let variableDecl = members.compactMap {
  $0.decl.as(VariableDeclSyntax.self) } // 2
let variablesName = variableDecl.compactMap {
  $0.bindings.first?.pattern } // 3
let variablesType = variableDecl.compactMap {
  $0.bindings.first?.typeAnnotation?.type } // 4
```

Let's explain the preceding code, line by line:

1. We use the declaration parameter to get all the struct members.
2. All the struct members also include their functions, so we filter it only to variables.
3. We create an array of all the variable's names using their `pattern` attribute.
4. We create another variety with all the variable types, using their `typeAnnotation` attribute.

Now that we have all the information we need, we can generate our Swift code for the `init` function.

First, we generate the `init` function signature based on the list of variable names and types:

```
var code = "init("
for (name, type) in zip(variablesName, variablesType) {
    code += "\(name): \(type), "
}
code = String(code.dropLast(2))
code += ")"
```

The preceding code starts by creating a mutable string, looping all the variable names and types, and adding them to the function signature. Once the code adds all the function parameters, it closes with a closing parenthesis.

Next, it's time to add the function body. We can do that using a special `SwiftSyntax` struct that represents an initializer declaration called `InitializerDeclSyntax`:

```
let initializer = try InitializerDeclSyntax(SyntaxNodeString
  (stringLiteral: code)) {
      for name in variablesName {
          ExprSyntax("self.\(name) = \(name)")
      }
}
```

The `InitializerDeclSyntax` "init" function receives two parameters – the function signature and a closure with the "init" body represented by `ExprSyntax`.

Now that we have `initializer`, we can return an array of `DeclSyntax`:

```
return [DeclSyntax(initializer)]
```

Let's see the full code:

```
let members = structDecl.memberBlock.members
      let variableDecl = members.compactMap {
        $0.decl.as(VariableDeclSyntax.self) }
      let variablesName = variableDecl.compactMap {
        $0.bindings.first?.pattern }
      let variablesType = variableDecl.compactMap {
        $0.bindings.first?.typeAnnotation?.type }

      var code = "init("
      for (name, type) in zip(variablesName,
        variablesType) {
          code += "\(name): \(type), "
      }
      code = String(code.dropLast(2))
```

```
        code += ")"

        let initializer = try InitializerDeclSyntax(SyntaxNodeString
          (stringLiteral: code)) {
            for name in variablesName {
                ExprSyntax("self.\(name) = \(name)")
            }
        }

        return [DeclSyntax(initializer)]
```

The code takes the struct list of variables and generates its own `init` function.

How does it look? Let's demonstrate that with a small struct:

```
struct Book {
    var id: Int
    var title: String
}
```

The `expansion` method creates the following `init` function:

```
init(id: Int, title: String) {
    self.id = id
    self.title = title
}
```

But the fact that we just defined the macro behavior doesn't mean we can use it. Remember that the macro runs as a compiler plugin. That's our next step.

Adding the compiler plugin

The compiler plugin is our macro "product," or, in other words, the macro entry point.

In iOS, macros are invoked in a sandbox without network access and system file changes. The question is this: How does the compiler instantiate and store Swift macros to be used as a plugin?

The answer is that it doesn't. If we have another look at our code, we'll notice that Swift Macros functions are all static, and that's an important issue when creating a new macro.

So, to create a compiler plugin, we need to define a new struct that conforms to the `CompilerPlugin` protocol and has the `@main` attribute mark:

```
@main
struct struct_initial_macroPlugin: CompilerPlugin {
    let providingMacros: [Macro.Type] = [
        StructInit.self,
```

```
        ]
}
```

The preceding code shows that `struct_initial_macroPlugin` implements one variable get method – `providingMacros` – and returns an array of macro types instead of instances.

Another essential thing to notice here is the struct name (`struct_initial_macroPlugin`). It doesn't matter what name we give it as long as it conforms to the `CompilerPlugin` protocol and has the `@main` attribute.

Now that we have a compiler plugin, our compiler is ready to run it.

Running our macro using a client

Macro executables are different than apps or libraries since they run in a compiler environment. If we go back in our chapter to the section where we create the Swift Macros Swift package (*Examining our Swift Macros package structure* section), we see that the Swift macro has another folder called `StructInitClient`.

`StructInitClient` is our Swift macro executable, also defined in the macro's `package.swift` manifest file:

```
.executable(
    name: "StructInitClient",
    targets: ["StructInitClient"]
),
```

Now, we can change the code we have in the `main.swift` file to the following:

```
import StructInit
import Foundation

@StructInit
struct Book {
    var id: Int
    var title: String
    var subtitle: String
    var description: String
    var author: String
}
```

In the preceding code, we have a simple struct named `Book`, but now, we have also attached the `@StructInit` macro we just created.

Right-click on the macro itself and choose **Expand Macro**, which reveals the generated code (*Figure 10.8*):

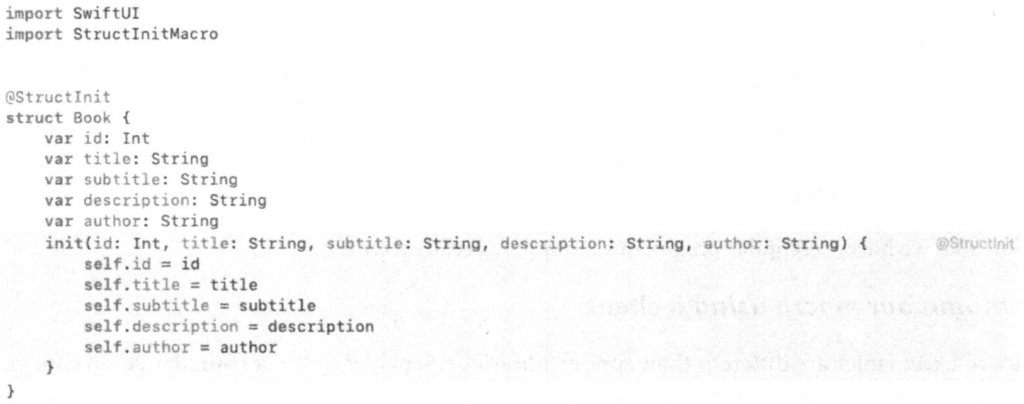

```swift
import SwiftUI
import StructInitMacro

@StructInit
struct Book {
    var id: Int
    var title: String
    var subtitle: String
    var description: String
    var author: String
    init(id: Int, title: String, subtitle: String, description: String, author: String) {      @StructInit
        self.id = id
        self.title = title
        self.subtitle = subtitle
        self.description = description
        self.author = author
    }
}
```

Figure 10.8: Swift macro expansion

Using our macro executable is a great way to see our macro in action! At this point, everything should work as expected. It's time to level up our macro implementation with some error handling.

Handling macros errors

When we create a Swift macro, things obvious to us, as the macro developers, are not obvious to our macro users.

Our `StructInit` macro is designed to function exclusively with structs, not classes. Therefore, we need to check whether the attached element is indeed a struct.

Inside the `expansion()` function, we can perform a simple `guard` statement and throw an error in case the attached declaration is not a struct:

```swift
guard let structDecl = declaration.as(StructDeclSyntax.self)
    else {
      throw StructInitError.onlyStructs
    }
```

In the preceding code, `StructInitError` is an enum that conforms to `Error`:

```swift
enum StructInitError: CustomStringConvertible, Error {
    case onlyStructs
    
    var description: String {
        switch self {
```

```
            case .onlyStructs: return "@StructInit can only be applied to
a structure"
        }
    }
}
```

Having an enum with different error types and messages can make a developer's life much easier. Remember that this error appears in compile time (*Figure 10.9*):

```
@StructInit                          ⊗  @StructInit can only be applied to a structure
class BookClass {
    var id: Int = 0
}
```

Figure 10.9: An error message is thrown when implementing a Swift macro

But sometimes, we want to handle more complex errors. For example, sometimes we want to show a warning, not just an error. Or, in other cases, we even want to offer our developer a fix for their problem.

In these cases, we can add something called a `Diagnostic` struct. A `Diagnostic` struct is more suitable for showing errors in a compiler environment and has more capabilities than just throwing errors.

Let's create a `DiagnosticMessage` enum and a `Diagnostic` struct:

```
enum CustomDiagnostic: String, DiagnosticMessage {
    case notAStruct

    var severity: DiagnosticSeverity { return .error}
    var message: String {
        switch self {
        case .notAStruct:
            return "@StructInit can only be applied to a structure"
        }
    }
    var diagnosticID: MessageID {
        return MessageID(domain: "StructInitMacro",
                         id: rawValue)
    }
}
let diagnostic = Diagnostic(node: node,
    message: CustomDiagnostic.notAStruct)
```

The code is much longer now! However, it contains much more information and features and is also structured in a way built for the `SwiftSyntax` library.

If you wondered why we need the `context` parameter in the `expansion` function, now you'll have the answer:

```
context.diagnose(diagnostic)
```

Remember we said that context links us to the compiler environment? So, we use it to invoke a diagnostic message.

Let's see the `guard` declaration now that we have a `diagnostic` structure:

```
guard let structDecl = declaration.as(StructDeclSyntax.self) else {
        let diagnostic = Diagnostic(node: node,
                                    message: MyLibDiagnostic.notAStruct)
        context.diagnose(diagnostic)
        throw StructInitError.onlyAStruct
}
```

We can see that `SwiftSyntax` is like peeling an onion – we uncover new features every time we dig deeper, and `Diagnostic` is one of these features.

Now, we have a significant error handling – descriptive and precise. But what about checking our macro in various use cases?

To see our macro at work, we used `StructInitClient`. However, relying on the client to verify that our macro works as expected is not sustainable over time.

So, another great feature we get from having a macro written in a Swift package is unit tests.

Let's see how we test a macro.

Adding tests

The principle of testing a macro is to test a code block *before and after* the macro expansion.

As part of our Swift Macros package, we have a test target (*Figure 10.10*):

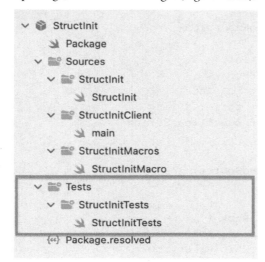

Figure 10.10: A testing target for StructInitMacro

Each Swift package comes with a testing target, and in this case, we already have one test with the `stringify` macro that comes when we create a new Swift Macros package.

Let's clear the test file and start our own test.

To test a macro, we need to create the `XCTestCase` subclass and create a new method called `testMacro`. Remember that test functions in `XCTest` always start with the phrase "test" followed by the test name.

To test a macro expansion, we will use a particular `SwiftSyntax` function called `assertMacroExpansion`. The most important function parameters are as follows:

- `_originalSource`: The original code before the expansion, including the macro attribute itself
- `expandedSource`: The code *after* the expansion
- `macros`: The list of macro types being tested

Let's see a basic test case for testing our `StructInit` macro:

```
let testMacros: [String: Macro.Type] = [
    "StructInit": StructInit.self,
]
```

```swift
final class StructInitTests: XCTestCase {
    func testMacro() {
        assertMacroExpansion(
            """
            @StructInit
            struct Book {
                var id: Int
                var title: String
                var subtitle: String
            }
            """,
            expandedSource:
            """

            struct Book {
                var id: Int
                var title: String
                var subtitle: String
                init(id: Int, title: String,
                  subtitle: String) {
                    self.id = id
                    self.title = title
                    self.subtitle = subtitle
                }
            }
            """,
            macros: testMacros
        )
    }
}
```

We can see that `assertMacroExpansion` received the three parameters I mentioned earlier.

We compare the `Book` struct expansion with the `Book` struct desired structure, including the `init` function.

`assertMacroExpansion` compares the expanded code of the macro to the `expandedSource` parameter, and if there are any differences, it fails the test.

Testing is a crucial part of Swift packages in general. Swift packages are meant to be reusable and rely on testing to ensure their stability.

Things get even more important when creating Swift macros since they run as a compiler plugin, which makes it harder to debug. So, we shouldn't give up tests, especially not in macros.

Practice exercises

Swift Macros is a complex topic, and it is a challenge to understand how to create a Swift macro without trying it yourself. Here are two exercises that can help you get started:

- Create an attached Swift macro that adds a function called `printVariables`. The function prints the list of the class properties and their values.
- Create a freestanding macro called `#colorhex` that receives a hex color value and generates an RGB color expression. For example, `#colorhex("#FFFFFF")` will generate `Color(red: 0.0, green: 0.0, blue: 0.0)`.

In addition, here are some links that can help you get more insights about Swift Macros:

- **Swift Macros documentation from the Swift.org projects**: `https://docs.swift.org/swift-book/documentation/the-swift-programming-language/macros/`
- **A GitHub repository about great Swift macros we can use and learn from**: `https://github.com/krzysztofzablocki/Swift-Macros`

Summary

This chapter covered a new and exciting feature of Xcode 15 and iOS 17 – Swift Macros.

We explored the `SwiftSyntax` library and learned how to set up, parse, and generate Swift code. We also created our first Swift macro, handled errors, and even wrote one test.

Swift Macros is a comprehensive, complex, yet effective feature, and by now, you are ready to implement it in your own projects!

In the next chapter, we'll discuss another exciting framework – Combine.

11
Creating Pipelines with Combine

Data flow is a central programming topic, not just in iOS development. Indeed, we have many solutions and design patterns to address data flow management. It was only in 1997 that the computer science world introduced reactive programming – a programming paradigm focusing on data streams, enabling declarative composition.

Apple's version of reactive programming is Combine, a framework that provides infrastructure for building data streams in our apps. It is also the infrastructure of SwiftUI, enabling it to be a declarative framework.

In this chapter, we will do the following:

- Discuss the reasons to use Combine in our projects
- Go over the basics
- Delve into Combine
- Learn about Combine using examples

Before we start going over the Combine framework, let's understand why we should use Combine.

Technical requirements

For this chapter, it's essential to download Xcode version 16.0 or higher from the App Store.

Ensure you're operating on the most recent version of macOS (Ventura or newer). Just search for Xcode in the App Store, choose the latest version, and proceed with the download. Open Xcode and complete any further setup instructions that appear. After Xcode is completely up and running, you can begin.

Download the sample code from the following GitHub link: `https://github.com/PacktPublishing/Mastering-iOS-17-Programming-fifth-edition/tree/main/Chapter%2011`

Why use Combine?

Apple's Combine framework is considered to have a steep learning curve, but not because it is technically complex. This is because many developers don't understand why, how, and where they should use Combine in their apps.

To answer these questions, let's try to understand Combine. Combine is Apple's reactive framework and provides a unified API for asynchronous events and data streams.

But why do we need a reactive framework? Don't we have everything we need?

Let's see what we have in our iOS SDK:

- **Notifications** allow us to send messages that any object can observe
- **Delegates** allow objects to respond to events or changes triggered by other objects
- **Closures** are self-contained functionality blocks we can pass around and call whenever we need
- **Key-Value Observing** (**KVO**) allows us to observe value changes in object property

That's a powerful toolbox! But while we have a toolbox with so many options, these options have some drawbacks we need to discuss.

For example, notifications might be considered to be an anti-pattern, mainly because they have implicit communication. Imagine a project based on notifications, with objects that mainly communicate with each other using the notification center. That project can take time to manage and understand the data flow and dependencies.

Delegates help with some of these problems. But when we want to pass data between different objects continuously, they require us to create many protocols and require each object to call another, making it hard to understand what's happening.

Closures are actually significant progress compared to delegates, but they also create complexity when nested or captured by other closures.

Imagine we have a view controller with a view model. The view model has a `message` property, and we always want UILabel's text to match the `message` property value.

With Combine, we would do the following:

```
messageSubscriber = viewModel.$message
        .sink { [weak self] message in
            self?.label.text = message
        }
```

The `sink` operator receives an update of any change in the `message` property and has a closure with the new `message` property. We store the new `message` property directly in the label's text value.

This example binds the label's text property to the `viewModel` text property. There is no need to define a specific interface for delegating, observing, posting notifications, or defining a closure.

Combine has much more to offer, but before we delve into additional practical examples, let's understand the basics.

Going over the basics

As a reactive framework, Combine is built upon components that publish updates (the publishers) and components that subscribe to updates (the subscribers).

In between, we've got the operators, which can manipulate data and control the stream flow. Let's get an overview of Combine by starting with the publisher.

Starting with the publisher

The best way to explain how Combine works is by talking about publishers. **Publishers** are types that can deliver a sequence of values over time. We saw one example in *Chapter 10*:

```
URLSession.shared.dataTaskPublisher(for: url)
```

For a type to be a publisher, it needs to conform to the `Publisher` protocol, and `URLSession` is not the only type that does that. `Timer` and `NotificationCenter` are also types that have their publishers:

```
NotificationCenter.default.publisher(for:
  Notification.Name("DataValueChanged"))
```

Or, it can be a `Timer` publisher:

```
let timerPublisher = Timer.publish(every: 1.0, on: .main,
  in: .default)
    .autoconnect()
```

For types that don't have a publisher, we can add one as long as their property is KVO-compliant:

```
extension UserDefaults {
@objc dynamic var test: Int { return integer(forKey:
  "myProperty") }
}
let userDefaultsPublisher = UserDefaults.standard
  .publisher(for: \.myProperty)
```

We can also create a custom publisher, and we'll learn how to do that shortly. The publisher emits values only if a subscriber wants to receive them. So next, let's meet the subscriber.

Setting up the subscriber

A **subscriber** is a protocol for instances that can receive values from a publisher. The `Subscriber` instance is located at the end of the stream and handles the incoming values. The Combine framework has two built-in subscribers, `sink` and `assign`; both simplify the use of Combine in most cases.

Let's start with `sink`:

```
import Combine
import Foundation

let subscriber = Timer.publish(every: 1.0, on: .main, in:
  .default)
    .autoconnect()
    .sink( receiveValue: { value in
        print("Received value: \(value)")
    })

DispatchQueue.main.asyncAfter(deadline: .now() + 5) {
    subscriber.cancel()
}
```

In this code example, we created a `Timer` publisher that sends a value every second. The value it sends is from the date type, but it doesn't matter to us – the `sink` subscriber can receive any value.

The next thing we do is cancel the subscriber after five seconds. Once no subscriber is listening, the publisher stops sending values. That's an essential concept of Combine, called a **demand-driven model**. With this approach, we ensure efficient resource management and avoid performing any work without a goal.

In this code example, we printed the received value to the console. However, in many cases, we want to assign it to a specific property. For example, we may download a file and receive an update on its progress. In this case, we want to update a progress property to show the download status.

We could use the `sink` closure to receive the value and set it to the relevant property, but we've got a more elegant way, and that's the `assign` subscriber:

```
import Combine
import Foundation

class DateContainer {
    var date: Date
```

```
        init() { date = Date() }
}

let container = DateContainer()

let cancellable = Timer.publish(every: 1.0, on: .main, in:
   .default)
      .autoconnect()
      .assign(to: \.date, on: container)

DispatchQueue.main.asyncAfter(deadline: .now() + 2) {
    cancellable.cancel()
}
```

In this example, we have the same timer as the previous example. However, this time, we've got an instance of `DateContainer` with a `date` property. The `assign` subscriber at the end of the stream ensures that we take the received value and assign it to a specific property using a key path.

In this case, the `assign` subscriber input value must match the publisher output value.

We obviously can achieve the same results using the `sink` closure:

```
      .sink( receiveValue: { value in
           container.date = value
      })
```

However, using the `assign` subscriber is far more elegant and more than just semantic. Using key paths improves our code type safety and makes it more declarative and concise.

We've learned that the publisher's output should match the subscriber's input. But what do we do if we need to perform some transformations and processing to make that happen? That's why we have the operators.

Connecting operators

The third part of Combine streams is the **operator**. The operator takes upstream data (the output from the previous step), processes it, and emits it downstream. Downstream means the next step – the subscriber or another operator.

The operators are actually what helps us build what we call a **pipeline** or **Combine stream**.

Let's try to build a simple stream:

```
let numbersPublisher = Array(1...20).publisher

let subscription = numbersPublisher
```

```
.filter { $0 % 2 == 0 }
.map { "The number is \($0)" }
.sink(receiveValue: { print($0) })
```

This simple code example takes an array of numbers between 1 and 20, making it a publisher using the `publisher` variable.

`numbersPublisher` emits a new value from the array each time. The value goes downstream to the `filter` operator, which republishes the value only if it's even. The filtered value moves to the `map` operator, which transforms it into a string message and republishes it again.

At the end of the stream, we have the `sink` subscriber, which prints the message to the console.

Congratulations! We've created our first pipeline. Look at *Figure 11.1*:

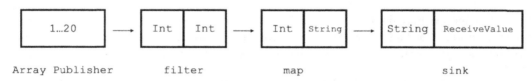

Figure 11.1: Our first Combine timeline

Figure 11.1 shows the different pipeline operators, such as `filter` and `map`. I am highlighting the input and output for each operator here. We can see that the output of one timeline component is the input of the next element.

That leads us to something we haven't discussed yet – what exactly are publishers and subscribers? How do they work under the hood? Let's delve in.

Delving into Combine components

Until now, we have created simple examples in Combine to warm up. However, if we want to use Combine in a more advanced way, we need to understand better what happens under the hood.

The first thing we must understand is that Combine is not magic. Combine alone doesn't include any sophisticated code. Ultimately, we are talking about a group of protocols that helps us subscribe to changes and create a pipeline of updates.

To delve in, we will review the different protocols and build our own custom publishers, operators, and subscribers to understand how things work inside.

Let's start with the publisher.

Creating a custom publisher

I just mentioned that Combine is a set of protocols that speak with each other, and the publisher is the first protocol we will review.

Let's see what we know up until now about the publisher:

- The publisher *emits values* to one or more subscribers
- The publisher output type *must match the subscriber's input*
- The publisher can also *deliver errors*

Based on that, let's take our `Int` array publisher example and try to create our own publisher that delivers numbers:

```
class CustomNumberPublisher: Publisher {
    typealias Output = Int
    typealias Failure = Never

    private let numbers: [Int]

    init(numbers: [Int]) {
        self.numbers = numbers
    }

    func receive<S: Subscriber>(subscriber: S) where
      S.Input == Output, S.Failure == Failure {
        for number in numbers {
            _ = subscriber.receive(number)
        }
        subscriber.receive(completion: .finished)
    }
}
```

The `CustomNumberPublisher` class has three essential parts:

- `Output` – This is where we define the publisher output type. In this case, it is an `Int` type.
- `Failure` – This is where we define the publisher error type. In this case, the publisher never emits an error.
- `receive` – This is the main publisher method. Combine calls the `receive` method whenever a subscriber subscribes to the publisher. We can see that the `receive` function has the subscriber's parameter, and it also verifies that the subscriber input type and error match the publisher definition.

When the publisher wants to emit a new value, it calls the subscriber's `receive` method with the new value. When the publisher completes sending values, it calls the subscriber's `receive` function with the `completion` parameter.

Let's see how we use `CustomNumberPublisher`:

```
let subscriber = CustomNumberPublisher(numbers: [1, 2, 3,
    4, 5])
    .sink { value in
        print(value)
    }
```

Running this code will print 1, 2, 3, 4, 5 to the console as expected.

The `CustomNumberPublisher` example explains how a publisher works. But sometimes, we want to send values imperatively. We may wish to implement Combine in an existing project code or simplify things.

So, let's meet a special publisher type called `Subject`.

Working with Subjects

A **Subject** is a publisher we can use to send values into a Combine stream. We do that by calling its `send(_:)` method.

Let's start with the most basic Subject – `PassthroughSubject`.

Understanding PassthroughSubject

Let's see a basic example of Subject usage:

```
import Combine

let subject = PassthroughSubject<Int, Never>()

let subscriber = subject.sink { value in
    print("Received value: \(value)")
}

subject.send(1)
subject.send(2)
subject.send(3)
```

The code example is simple and easy to follow. We created a `Subject` instance (which is a publisher) of the `PassthroughSubject` type. `PassthroughSubject` can be initialized without any value, and the first time we open a stream is after we call its `send(_:)` function.

Notice that our Subject is just sending values but never closing the stream. However, we've learned from our custom publisher implementation that, sometimes, the publisher closes its streams and sends a completion to the subscriber.

We can also use the send(_:) function to close the Combine stream:

```
subject.send(1)
subject.send(2)
subject.send(completion: .finished)
subject.send(3)
```

In this code example, we use our Subject to send two values – 1 and 2. After sending these values, we close the stream by calling the send function with the .finished parameter.

After that, the Subject tries to send another value (3), but the stream is already closed, and the subscriber won't receive it.

The publisher life cycle is crucial to the Combine methodology and applies to our Subjects.

PassthroughSubject is excellent for sending values to subscribers. However, it's not very good for holding a state. For example, imagine we want to store the current authentication login status or a file download progress. One solution is to store the received value in an instance variable. However, using an instance variable can be cumbersome, especially with several subscribers.

Another option is to use another type of Subject called CurrentValueSubject.

Preserving state with CurrentValueSubject Subject

Unlike PassthroughSubject, CurrentValueSubject is excellent for holding a state. It has an initial state and a value property representing the current value.

Let's see a basic example of CurrentValueSubject usage:

```
import Combine

let subject = CurrentValueSubject<String, Never>("Initial
  Value")

let currentValue = subject.value
print("Current value: \(currentValue)")

let subscriber = subject.sink { value in
    print("Received value: \(value)")
}

subject.send("New Value")
```

In this code example, we create `CurrentValueSubject` and initialize it with a value (`"Initial Value"`).

We then print the Subject's current value into the console and subscribe to it using a simple `sink` function, printing each update as well.

In the last line, we send a new value using our Subject.

The console, in this case, will show the following:

```
Current value: Initial Value
Received value: Initial Value
Received value: New Value
```

At first glance, the console output looks weird. Why do we see `Received value: Initial Value` if we have not sent it using the `send(_:)` function?

The answer is that `CurrentValueSubject` already holds a value when we initialize it, and when we subscribe to it for the first time, we already receive the current value.

This is why `CurrentValueSubject` is excellent for state management. This behavior ensures that our subscribers always sync with the current Subject value.

`PassthroughSubject` doesn't have the `value` property, and we cannot read its current value. However, the fact that it doesn't emit its values before we call the `send(_:)` function can be an advantage in some cases.

Let's see an example:

```
let subject1 = PassthroughSubject<Int, Never>()
let subject2 = PassthroughSubject<Int, Never>()

let subscriber = subject1
    .merge(with: subject2)
    .sink { value in
        print("Transformed value: \(value)")
    }

subject1.send(1)
subject1.send(2)
subject2.send(3)
subject2.send(4)
```

In this example, we have two `PassthroughSubject` publishers and the `merge()` operator in our Combine stream. The `merge()` operator combines the values emitted by both publishers into a single stream. If one of the Subjects sends a value, the `merge()` operator moves it down the stream.

So, in this case, the output will be as follows:

```
Transformed value: 1
Transformed value: 2
Transformed value: 3
Transformed value: 4
```

`PassthroughSubject` can act as an intermediate step in the Combine pipeline, allowing us to combine multiple publishers and perform data transformations before it reaches the subscribers. This is something we cannot do with `CurrentValueSubject`.

Until now, we have used the built-in `sink` subscriber to handle the incoming value. But just like the publisher, we can also create a custom subscriber. Learning how to make a custom subscriber can enrich our knowledge of Combine. Let's delve in!

Creating a custom subscriber

If the publisher is the element that delivers updates, the subscriber is the element that demands them.

We already understand that Combine works with a *supply-and-demand* model. This means the subscriber needs a mechanism to handle and request incoming values.

Let's build a subscriber for `CustomNumberPublisher`:

```
class CustomNumberSubscriber: Subscriber {
    typealias Input = Int
    typealias Failure = Never

    func receive(subscription: Subscription) {
        subscription.request(.unlimited)
    }

    func receive(_ input: Int) -> Subscribers.Demand {
        print("Received: \(input)")
        return .unlimited
    }

    func receive(completion: Subscribers.Completion<Never>)
    {
        print("Received completion: \(completion)")
    }
}
```

The subscriber protocol contains the publisher's corresponding type aliases, `Input` and `Failure`. Both need to match the publisher's `Output` and `Failure` data types.

Looking at the subscriber implementation, we can see three more `receive` functions, which we'll look at in the next subsections.

receive(subscription: Subscription)

`receive(subscription: Subscription)` is called when the subscriber subscribes successfully to the publisher. The `subscription` object handles the subscription, and it has one important method – to define the demand from the publisher. We do that by requesting unlimited values:

```
subscription.request(.unlimited)
```

We can also limit the number of items we're requesting. For example, let's request a maximum of three additional items:

```
subscription.request(.max(3))
```

We can also request no items at all:

```
subscription.request(.none)
```

Notice that the publisher needs to call the `receive(subscription: Subscription)` method explicitly. This means that if we build a custom publisher (as in the *Creating a custom publisher* section), we must ensure that we call that function ourselves.

We need to handle the incoming values now that we have established the subscription, and we do that with the `receive(_input:Int)` method.

receive(_ input: Int) -> Subscribers.Demand

If we look back at the `CustomNumberPublisher` subscriber we created in the *Creating a custom publisher* section, we can see that our publisher calls the subscriber directly:

```
_ = subscriber.receive(number)
```

That's the `receive(_ input:Int)` method we need to implement as part of the subscriber protocol. This method handles the incoming updates, similar to the closure we saw in the `sink` function (in the *Setting up the Subscriber* section).

Notice that the `receive` function returns `Subscribers.Demand`. That's the same demand type we discussed in the previous function. When the subscriber finishes handling the input, it must inform the publisher how many more items it demands. Demanding more items doesn't replace the demand sent in the previous function when the subscriber first established its subscription to the publisher. The new demand request is an additive value that needs to be handled by the publisher.

Look at the following code:

```
func receive(subscription: Subscription) {
    subscription.request(.max(2))
}

func receive(_ input: Int) -> Subscribers.Demand {
    print("Received: \(input)")
    return .max(3)
}
```

Let's try to follow the calls in this code example:

- The subscriber subscribes to the publisher, and the `receive(subscription: Subscription)` function is called, returning a maximum of 1. The total demand is now 1.
- The publisher emits a value to the subscriber, and the `receive(_ input:Int)` function is called, returning a maximum of 3. The total demand is now 4.

As mentioned, it is the publisher's responsibility to manage the subscriber demand. If we create our custom publisher, we need to consider that.

Now that we know how to start and manage a subscription, it's no less important to understand how to complete it.

receive(completion: Subscribers.Completion<Never>)

The publisher calls the subscriber's `receive(completion:)` function when it completes publishing. It can be either because the publisher has no updates or an error.

That's where the subscriber needs to perform cleanups, update UI or application state, or print logs, mainly when an error occurs.

Here's an example of the `receive(completion:)` function:

```
func receive(completion: Subscribers.Completion<Never>) {
    switch completion {
    case .finished:
        print("Subscription completed successfully.")
    case .failure(let error):
        print("Subscription failed with error: \(error)")
    }
}
```

That's a basic implementation of the `receive(completion:)` function.

We now know how to create a custom publisher and a custom subscriber. Now, let's see how to connect them.

Connecting the custom publisher and subscriber

To complete the picture of how the subscriber and the publisher work together, we must return to the publisher and respond to our subscriber demand requests.

Let's see an example of how to implement a `receive` function on the publisher side:

```
func receive<S: Subscriber>(subscriber: S) where
  S.Input == Output, S.Failure == Failure {
  for number in numbers {
      guard subscriber.receive(number) != .none else
      {
          subscriber.receive(completion: .finished)
          return
      }
  }
  subscriber.receive(completion: .finished)
}
```

In the code example, the publisher keeps sending more updates to the subscriber as long as the subscriber keeps demanding them. When the subscriber stops demanding more updates, the publisher closes the stream and calls the subscriber `receive(completion:)` function.

At this point, we should be familiar with how subscribers and publishers work together. We created custom publishers and subscribers and performed basic subscriptions. Let's improve these subscriptions with operators, something we have barely discussed.

Working with operators

Subscriptions and publishers are great, but the true power of Combine comes from operators.

Unlike subscriptions and publishers, operators are not protocols or instances. Operators are just publisher methods that republish the update downstream and create a chain of data manipulations until the subscriber reaches the end of the pipeline.

Operators help us modify the updates, filter them, merge them, and perform many operations, which allows us to achieve an ideal result.

The Combine framework comes with many built-in operators. We will go over only some of them now, but you can go over the full list at the Apple website: https://developer.apple.com/documentation/combine/publishers-catch-publisher-operators

Let's start with some basic operators.

Starting with basic operators

One of the most basic use cases for operators in Combine is to *filter* the updates that the publisher delivers.

For example, we can use the `filter` operator:

```
let cancellable = (1...10).publisher
    .filter{ $0 % 2 == 0 }
    .sink { value in
        print(value)
    }
```

In this code example, we created a publisher that emits values from 1 to 10. The `filter` operator ensures that only even numbers will continue downstream. This code will print 2, 4, 6, 8, and 10 to the console.

Another example of a filtering operator is `removeDuplicates`:

```
let cancellable = [1,2,2,3,3,3,4,5].publisher
    .removeDuplicates()
    .sink { value in
    print(value)
}
```

The code example shows a publisher that emits duplicate values. The `removeDuplicates` operator filters out the values in case they were sent in the last update. The console will show the following:

```
1
2
3
4
5
```

Let's try to create a custom operator to understand how an operator works underneath.

Creating a custom operator

When we try to examine the filter operator in Apple's header files, we can see the following:

```
extension Publisher {

...
public func filter(_ isIncluded: @escaping (Self.Output) ->
   Bool) -> Publishers.Filter<Self>
}
```

`filter()` is a function that accepts a closure with a parameter of a generic type of `Output` and returns a publisher. This function extends the `Publisher` protocol we discussed earlier under *Creating a custom publisher*.

The important thing to notice here is that the `filter` function republishes the values and allows multiple operators to be chained together to create a complex data processing pipeline.

This is similar to how view modifiers work with SwiftUI – they modify the current view and return a new view.

To create our own custom operator, let's try to do the same thing and create a `multiply` operator. Our `multiply` operator accepts an `Int` value and republishes it while multiplying with a certain factor:

```
extension Publisher where Output == Int {
    func multiply(by factor: Int) -> Publishers.Map<Self,
      Int> {
        return self.map { value in
            return value * factor
        }
    }
}
```

In our code example, we also extended the `Publisher` protocol while ensuring the `Output` type needs to be `Int`.

We then create a `multiply` function that accepts a factor as a parameter and returns a new publisher.

In our implementation, we use a `map` operator to transform our value into a new one, which means we need to return a `Map` publisher. Let's see how to use our new operator:

```
let cancellable = [1, 2, 3, 4, 5].publisher
    .multiply(by: 2)
    .sink { value in
        print("Received value: \(value)")
    }
```

We added our new `multiply` operator to a Combine stream that starts with an array of five numbers. The output for this code would be as follows:

```
Received value: 2
Received value: 4
Received value: 6
Received value: 8
Received value: 10
```

We created our first operator!

However, if you are like me, the return of a new Map publisher may bother you. Let's try to understand why it happened and what we can do about it.

Working with AnyPublisher

Our intuition says that if `multiply` is a function that accepts an `Int` type and returns a new value, why do we need to use a Map publisher?

So, we need to remember that operators republish our values. The function doesn't return a value but rather a publisher that publishes the new value. It might not sound obvious, but our goal is to create a chain of publishers and multiply, despite its name, which is part of this chain.

So, our solution is to return some sort of a *generic publisher*, or what we call in Combine – `AnyPublisher`.

`AnyPublisher` is a type-erased publisher, and we use it to present a more abstract interface to our publishers.

Let's see our `multiply` operator version, which now returns `AnyPublisher` instead of `Publisher.Map`:

```
extension Publisher where Output == Int {
    func multiply(by factor: Int) -> AnyPublisher<Int,
      Failure> {
        return self.map { value in
            return value * factor
        }
        .eraseToAnyPublisher()
    }
}
```

In this code example, we performed two changes:

- We changed the function's return type to `AnyPublisher<Int, Failure>`. In this way, we hide the implementation details and the fact that we used the Map publisher.
- We erased the publisher type using the `eraseToAnyPublisher()` function, which erases the publisher type and returns `AnyPublisher`.

At first glance, it looks like `AnyPublisher` is there only for semantic reasons. But when I said that returning a Map publisher *bothers* me, it wasn't because it didn't look nice. It's because `AnyPublisher` has practical implications for how we build Combine streams.

One reason is *API design*. Using `AnyPublisher` allows us to design a more flexible and polymorphic API interface. Our previous version of the `multiply` function returned a specific type of publisher. Returning `AnyPublisher` makes it easier to chain publishers together because they are from the same type.

Another reason is *decoupling* – by returning publishers as `AnyPublisher`, we're decoupling our publisher's implementation from its usage. By that, we make our code more modular and maintainable.

The `filter` and `removeDuplicates` operators, along with `map`, are great for streamlining and manipulating values along the pipeline. We also reviewed the `merge` operator when we discussed Subjects in the *Working with Subjects* section. But Combine offers more advanced operators. Let's go over some of them now.

Exploring advanced operators

Let's face it, up until now, we discussed operators that performed tasks that were easy to do even without Combine. Yes, using `map` and `filter` operators is extremely valuable, but they don't reflect the real Combine added value.

One of the Combine framework goals is to create much more sophisticated and complex streams that can be error-prone to do without it.

Let's understand what I mean and explore the `zip` operator.

Using the zip operator

The `zip` operator combines values from two publishers and emits a tuple only after each publisher emits its value.

Once the `zip` operator receives values from all publishers, it emits a tuple and *resets* itself. This means it waits again to receive values from all publishers before it emits a new tuple.

Let's see a simple code example:

```
import Combine

let publisher1 = PassthroughSubject<Int, Never>()
let publisher2 = PassthroughSubject<Int, Never>()

let cancellable = publisher1
    .zip(publisher2)
    .sink { value in
    print("Zipped value: \(value)")
}

publisher1.send(1)   // no output
publisher2.send(10)  // output is (1,10)
publisher1.send(2)   // no output
publisher2.send(20)  // output is (2,20)
```

In this code example, we used two Subjects to send values to our subscriber. We zipped them together and printed the output.

We can see that after `publisher1` sends a value, the stream doesn't continue and waits for `publisher2` to send its value. Only after `publisher2` sends a value does the stream continue and print (1, 10) to the console. At this point, the `zip` operator is reset, and again, it waits for both publishers to emit values.

The `zip` operator is not limited to two publishers. We can also use `zip` for three publishers and receive a tuple of three.

The `zip` operator belongs to a group of Combine operators that handles multiple publishers together. We already saw the `merge` operator under the *Working with Subjects* section.

Another operator that belongs to this category is `combineLatest`. Let's go over it now.

Combining multiple values using combineLatest

The `zip` operator combines multiple publisher outputs into a tuple. However, it waits for all publishers to send values each time.

The `combineLatest` operator only waits for the first time for all publishers to emit values, and from this point, it emits a new tuple each time one of the publishers sends a new value.

Let's see an example of `combineLatest`:

```
let publisher1 = PassthroughSubject<Int, Never>()
let publisher2 = PassthroughSubject<Int, Never>()

let cancellable = publisher1
    .combineLatest(publisher2)
    .sink { value in
    print("Combined value: \(value)")
}

publisher1.send(1)   // no output
publisher2.send(10)  // output will be 1,10
publisher1.send(2)   // output will be 2,10
publisher2.send(20)  // output will be 2,20
```

In this code example, we also have two Subjects that send values. This time, we combined them using `combineLatest`.

After `publisher1` sends its first value, `combineLatest` halts the stream as it waits for `publisher2` to send a value.

Once `publisher2` sends its first value, `combineLatest` emits a tuple with the values of (1, 10).

Next, `publisher1` sends a new value – 2. This time, `combineLatest` doesn't wait for `publisher2` to send a new value and emits a new tuple – (2, 10)

The behavior of emitting a new tuple each time one of the publishers sends a new value makes `combineLatest` a top-rated operator for handling asynchronized operations.

Imagine you have a screen being updated by multiple sources, such as a search results screen of live sports updates, and each time we get a new update, we want our screen to refresh its UI to reflect the new state.

`combineLatest` is ideal for such a case, as it creates a new tuple downstream whenever one of the publishers emits a new value.

We can use many more useful operators; you can find them all on Apple's website. However, the real challenge with adopting Combine in our projects is understanding how to implement them in real-life use cases.

Learning about Combine using examples

Up until now, we have discussed several Combine components and delved into understanding how Combine works underneath by creating our custom publishers, subscribers, and operators.

Despite that, many developers need help incorporating Combine frameworks in real-life scenarios.

The different publishers and operators are mostly clear in theory, but it can be difficult to imagine them as part of the central design patterns we use in our projects.

Let's review some examples to help us understand how to implement Combine in our projects. We'll start with a basic example of managing a UI state in a view model.

Managing UIKit-based view state in a view model

SwiftUI view states are naturally declarative. This means we can bind the view state, such as a list of items, to a UI component, such as a `List` view. That's the only way to handle states in SwiftUI.

However, achieving that design pattern in UIKit takes time and effort.

Using Combine, we can create a publisher and bind our table view data source to reflect any changes coming from the server.

Here's a code example for such a view model:

```
class MyViewModel {
    struct Item: Codable {
        let title: String
```

```
        let description: String
    }

    var dataPublisher: AnyPublisher<[Item], Error> {
        return URLSession.shared.dataTaskPublisher(for:
            URL(string: "https://api.example.com/data")!)
            .map { $0.data }
            .decode(type: [Item].self, decoder:
              JSONDecoder())
            .eraseToAnyPublisher()
    }
}
```

We have already discussed the `AnyPublisher` form, and that's a great example of its usage. We create a publisher that starts with a URL request, extracts its data using the `map` operator, and decodes it into an array of Items. To hide the publisher implementation, we erase its type for `AnyPublisher`. Connecting the view model to the view controller is simple now that we have a publisher:

```
viewModel.dataPublisher
    .sink(receiveCompletion: { completion in
    }, receiveValue: { [weak self] data in
        self?.updateTableView(with: data)
    })
    .store(in: &cancellables)
```

In this code example, we subscribe to our new `dataPublisher` and update our table view with the data.

To make our project even more modular, we can move the `URLSession dataTaskPublisher` function to a class of its own and keep the separation of concerns principle.

Performing searches from multiple sources

One of the most popular use cases of iOS development is performing a search from the server and the local database.

The requirement for such a search is first to show results from the local data store and then go to the server and return results.

This is a common requirement, and using Combine is also easy. Let's see a simple example of that:

```
func searchLocalDatabase(query: String) ->
AnyPublisher<[SearchResult], Never> {
    return Just([
        SearchResult(id: 1, title: "Local Result 1"),
```

```
            SearchResult(id: 2, title: "Local Result 2")
        ])
        .delay(for: .seconds(1), scheduler: DispatchQueue.main)
        .eraseToAnyPublisher()
}

func searchServer(query: String) ->
    AnyPublisher<[SearchResult], Never> {
    return Future { promise in
        DispatchQueue.global().asyncAfter(deadline: .now()
            + 2) {
            promise(.success([
                SearchResult(id: 3, title: "Server Result
                    1"),
                SearchResult(id: 4, title: "Server Result
                    2")
            ]))
        }
    }
    .eraseToAnyPublisher()
}

var cancellables = Set<AnyCancellable>()
let query = "example"
var totalResults = [SearchResult]()
searchLocalDatabase(query: query)
    .merge(with: searchServer(query: query))
    .sink(receiveCompletion: { _ in }, receiveValue: {
        results in
            totalResults.append(contentsOf: results)
            print("Search results: \(totalResults)")
    })
    .store(in: &cancellables)
```

In this code example, we performed three primary steps:

1. **We created a publisher for each source** – If we're working with several sources that need to deliver values (in this case, the local data source and the server), creating a publisher for each is useful. Once we have multiple publishers, combining them to create a stream of updates is easy. We used two publishers to simulate this search – Just and Future. We can use Just to start a stream, and Future is a publisher we use to perform a task and emit the value asynchronously.

2. **We used the merge operator to receive updates** – Now that we have a publisher for each source, we have merged them utilizing the `merge` operator. Remember that the `merge` operator emits an update if one of its sources emits a new value. We could also use `combineLatest`, but `combineLatest` waits for all the publishers to emit values before it emits the combined value downstream.

3. **We collected the data received** – Each time we got a new value, we appended its contents to the `totalResults` array. Our data flow doesn't have to end here. We can make `totalResults` a `CurrentValueSubject` instance and deliver the results to the view model or the view itself. If we are working with SwiftUI, we can make `totalResults` a `@Published` variable to refresh the search results UI automatically.

There's a nice lesson here related to using Combine in our projects. If we create publishers for different data sources and ensure they emit values, it becomes easy to create pipelines of updates and connect them to the rest of the project.

The following example handles another everyday use case, which is form validation.

Validating forms

Forms are common use cases in any user-facing platform, not just iOS. One of the most essential responsibilities of creating forms is the ability to validate their inputs.

Let's see how to use Combine to validate a simple sign-in form:

```
struct FormView: View {
    @ObservedObject var viewModel = FormViewModel()
    var body: some View {
        VStack {
            TextField("Username", text:
              $viewModel.username)
              .padding()
              .textFieldStyle(RoundedBorderTextFieldStyle())
            
            SecureField("Password", text:
              $viewModel.password)
              .padding()
              .textFieldStyle(RoundedBorderTextFieldStyle())
            
            Button("Login") {
                if viewModel.isFormValid {
                    print("Login successful!")
                } else {
                    print("Please fill in all fields.")
                }
            }
```

```
            }
            .padding()
            .disabled(!viewModel.isFormValid)
        }
        .padding()
    }
}
```

Our form contains two text fields – `username` and `password`. We also have a view model attached to the view. The view model has several `@Published` variables, such as `username`, `password`, and `isFormValid`. The `username` and `password` variables are connected to the view text fields.

Now, let's see the `FormViewModel` class:

```
class FormViewModel: ObservableObject {
    @Published var username: String = ""
    @Published var password: String = ""
    @Published var isFormValid: Bool = false
    
    private var cancellables = Set<AnyCancellable>()
    
    init() {
        Publishers.combineLatest($username, $password)
            .map { username, password in
                !username.isEmpty && !password.isEmpty
            }
            .assign(to: \.isFormValid, on: self)
            .store(in: &cancellables)
    }
}
```

When we initialize the view model, we create a Combine stream based on the `combineLatest` operator to observe changes in the `username` and `password` variables. The `map` operator ensures that both variables are not empty, and we assign the results (`Bool`) to the `isFormValid` variable.

The view observes the `isFormValid` value and uses it to turn the login button on and off.

This stream is basic; we can achieve the same results without Combine. However, forms can become very complex at some point. The Combine pipeline we created is an excellent infrastructure for more complicated forms.

Even simple rules for `username` and `password` can be easily enforced using our stream, as in this example:

```
Publishers.combineLatest($username, $password)
        .map { username, password in
            let isUsernameValid = !username.isEmpty &&
              username.count >= 6
            let isPasswordValid = !password.isEmpty &&
              password.count >= 8 && password.contains(
                where: { $0.isNumber })
            return isUsernameValid && isPasswordValid
        }
        .assign(to: \.isFormValid, on: self)
        .store(in: &cancellables)
```

In this code example, we used our Combine stream to enforce rules – the password needs at least eight characters including a number, and the username needs at least six characters. The `map` operator is great for centralizing this logic, outputting a Boolean value, and assigning it to the `isFormValid` value.

Summary

Combine makes our code reactive beyond SwiftUI views. It's a framework that can help us handle complex tasks such as search, network requests, and state management.

This chapter reviewed the basic Combine components, such as the publisher, subscriber, and operator. We also delved in and created custom versions of each of the components. We learned how to create pipelines with data transforms and network requests. In the end, we learned how to incorporate Combine in common use cases.

By now, we should be able to start working with Combine on our existing projects.

The next chapter touches on another topic many iOS developers feel irritated by – Core Data.

12
Being Smart with Apple Intelligence and ML

The launch of ChatGPT in November 2022 wasn't the first appearance of an **Artificial Intelligence** (**AI**) tool, but it was the one that put the AI in the spotlight.

Some may argue that Apple entered the AI world later than others. Perhaps, but what's certain is that iOS has machine-learning capabilities for both users and developers.

Machine learning opens up new capabilities in almost every area we can think of – from search, statistics, and insights to understanding images and sounds. There are even apps that are based on AI and machine learning capabilities.

Currently, most of these capabilities are server-based. Still, the ongoing improvements in mobile phones' **System On Chip** (**SoC**) performance allow them to perform predictions on-device, which opens up new opportunities.

In this chapter, we will do the following:

- Cover the basics of AI and machine learning, learn the different terms, how machine learning works, and what it means to train a model
- Explore built-in machine learning frameworks such as **Natural Language Processing** (**NLP**), vision, and sound analysis
- Add a semantic search to our Core Spotlight implementation
- Build and integrate a custom machine learning model using the Create **Machine Learning** (**ML**) application and the Core ML framework

Machine learning is a vast topic, and we've got much to cover, so let's jump right in to understand the basics.

Technical requirements

You must download Xcode version 16.0 or above for this chapter from Apple's App Store.

You'll also need to run the latest version of macOS (Ventura or above). Search for Xcode in the App Store and select and download the latest version. Launch Xcode and follow any additional installation instructions that your system may prompt you with. Once Xcode has fully launched, you're ready to go.

This chapter includes many code examples, some of which can be found in the following GitHub repository: https://github.com/PacktPublishing/Mastering-iOS-18-Development/tree/main/Chapter12.

Note that some examples in this chapter need to be run on a device, not the simulator.

Going over the basics of AI and machine learning

Before we dive in, let's acknowledge the complexity – AI and machine learning are two huge topics that are impossible to cover in one chapter or even one book.

However, it is recommended that we understand the basics if we want to implement some machine-learning capabilities in our projects.

So, let's start with understanding the difference between machine learning and AI.

Learning the differences between AI and machine learning

AI is considered a rising topic in computer science, and this trend has been accelerated since the launch of **ChatGPT**.

Even though ML is not a new technology, many people still need clarification about the difference between ML and AI. It's not that they are not related – they are. Still, as iOS professional developers, it is essential to have a clear overview of the differences now that Apple has integrated AI deeply into its system.

So, what is ML? ML technology focuses on developing algorithms and statistical models to help computers perform tasks such as prediction and classification. For example, a model can receive an image and reply whether it contains a cat, or a model can take some text and locate the verbs and the nouns.

The ML model is an algorithm that performs its predictions and classifications. In fact, a model can use several algorithms to perform its calculations. For example, a vision model can use neural networks to perform its image classification and a **YOLO (You Only Look Once)** algorithm to perform real-time object detection. Each algorithm has its strengths and weaknesses. For example, the decision tree algorithm is easy to interpret but is prone to overfitting, while **KNN (K-Nearest Neighbors)** is simple and intuitive but computationally intensive.

Conversely, AI is an array of technologies and methods that create a system capable of performing tasks similar to what humans usually do.

One great example is **LLM (Large Language Model)** services such as **ChatGPT** or **Gemini**. Another example is the autonomous car driving projects that involve many ML models, such as object detection and decision-making.

We now understand that the ML model is one building block in AI. Next, let's dive into the ML model to understand how it works.

Delving into the ML model

The **ML model** contains data to generate a prediction, classification, or decision. The ML models we want to store on a device are relatively small. However, models such as GPT can be hundreds of gigabytes.

But what does *data* mean here? How is it structured? The answer to that question depends on a model's algorithms. For example, if the model uses a linear regression algorithm, the data structure is a 2D array, where the rows represent samples and the columns represent features. A model with a **decision tree algorithm** contains a tree, where the leaves represent the different decisions or predictions.

Going forward in this chapter, we will refer to building and creating a model's data as training. Let's discuss that important topic.

Training the model

Two distinct ML models can have the same algorithms and structures, but the data can differ because of the training process. Using the training process, we teach a model to make accurate predictions and decisions based on the input data. This process involves optimizing the model's data (parameters) to accurately perform predictions on unseen data.

There are several steps we need to do to train a model:

1. **Data collection**: We need to prepare a relatively large dataset to train our model. We must also preprocess the data by handling missing values, cleaning unrelated data items, and normalizing values.
2. **Split the data collection**: Now that we have the dataset, we must divide it into training data, validation, and test sets. We use each of these sets in a different training stage.
3. **Pick our ML algorithm**: Each algorithm aims to solve a different problem. For example, the logistic regression algorithm solves classification problems, and the linear regression algorithm solves regression problems.
4. **Forward pass**: We pass the training data through the model to make predictions.
5. **Validation**: We use validation datasets to assess the model's performance and adjust the model based on the results.

6. **Testing**: We use the test data to evaluate our model's performance in real-time use cases with unseen input data.

That was a schematic overview of the training process. In practice, the process contains even more steps, such as calculating loss and optimizations. However, the goal is to give you a glimpse into training so that you can understand the following topics. And don't worry – we will build and train an ML model together soon!

Now that we know what an ML model is, let's try to understand how it relates to iOS.

Apple intelligence and ML

When ChatGPT gained popularity, many felt Apple had fallen behind in AI and ML. This book is not the place to discuss that question; suffice it to say that ML has since been an integral part of iOS for years. iOS uses ML to optimize our photos according to their content. Keyboard predictions involve ML models, and even the way iOS preserves a battery is based on ML.

All these features and capabilities are transparent to users and performed under the hood. However, iOS 18 brought AI into the spotlight with many features, such as an improved Siri, an image playground, and writing tools.

iOS 18 also provided some neat capabilities for us developers, but it especially brought the areas of ML and AI to our attention. For example, semantic search is one of the new capabilities available to developers using iOS ML features.

Before we dive into Core ML and learn how to train our models, let's start with the models that come with the iOS SDK, as there is a good chance that that is where we'll find what we need quickly without training a new model.

Exploring built-in ML frameworks

When we reviewed the basics of AI and ML, we saw what it means to train a model – it's a long and complex process. This process requires us to prepare relatively big datasets, including the validation and test datasets. Even after that, we have an ML model we need to fine-tune and include in our app while trying to reduce its size.

However, don't get me wrong. There are cases where training our ML model is essential, but before we start the training process, it's important to be familiar with what iOS offers.

Working with ML frameworks in iOS isn't new. These frameworks were introduced years ago, some even in iOS 10 (2016). However, few developers use them, perhaps because they believe they are complex to integrate.

We'll start with one of the most practical frameworks in the ML toolset – NLP.

Interpreting text using NLP

Interpreting and understanding texts can provide significant value to many apps. For example, NLP can help us understand strings such as search phrases, text inputs, or extracting information from an imported text.

The iOS SDK has a built-in NLP framework called `NaturalLanguage`:

```
import NaturalLanguage
```

The `NaturalLanguage` framework helps us interpret text efficiently on a device.

We must first know how NLP works under the hood and its basic terms to understand how it works.

The NLP model works by finding relationships between different parts of texts. Even though this task is complex, it's interesting to see how it works.

Understanding how NLP works

The NLP process involves text processing and several algorithms to extract the necessary information. There are three basic steps – **preprocessing**, **feature extraction**, and **modeling**. Let's go over them one by one.

Preprocessing

In this step, the NLP model starts by cleaning the input, such as removing duplicates, splitting texts into words and sentences, converting text to lowercase, and performing stemming and lemmatization. Take the following text as an example:

```
"Running is fun! I love to run."
```

This will be preprocessed to something like the following:

```
"run fun love run".
```

In this example, the NLP removed stop words (such as `is`) and lowercased the whole string.

Feature extraction

After the string has been preprocessed, we transform it into a feature set that we can use with the ML algorithm. In most cases, this involves capturing different patterns and word frequencies. For example, the string from the previous step, `run fun love run`, can be transformed into the following:

```
{
  "run": 2,
  "fun": 1,
```

```
    "love": 1
}
```

In this example, the NLP model takes the input string and analyzes the frequency of each word. This technique is called **Bag of Words (BoW)**, and the model uses it to determine the importance of the different words in the string. Note there are many feature extraction techniques, and BoW is just an example. We can select the model now that we have the feature extraction data.

Modeling

In the modeling step, we use string and feature extraction as input to the model algorithm. NLP uses several algorithms to analyze the string – logistic regression, naïve Bayes, and a neural network. The algorithm that the model selects depends on the task it needs to achieve.

For example, if the NLP framework needs to perform sentiment analysis, it would use a neural network-based model. Simple text processing tasks would use a rule-based system model.

These three steps demonstrate how complex it is to interpret a simple text. Fortunately, the `NaturalLanguage` framework performs all of these steps for us.

Let's see how to use the `NaturalLanguage` framework API.

Using the NaturalLanguage API

Finally, we are going to write some code! The `NaturalLanguage` framework has two primary uses – classification and word tagging. Let's start with classification.

Text classification

Using **text classification**, we can analyze the text sentiment to determine whether it is positive or negative.

For example, let's take a look at the following text:

```
The latest update made everything so much better. Great job!
```

To analyze the sentiment of this sentence using the `NaturalLanguage` framework, we'll use the NLTagger class:

```
let sentimentAnalyzer = NLTagger(tagSchemes:
  [.sentimentScore])
      sentimentAnalyzer.string = userInput

      let (sentiment, _) = sentimentAnalyzer.tag(at:
       userInput.startIndex, unit: .paragraph, scheme:
       .sentimentScore)
      if let sentiment = sentiment, let score =
       Double(sentiment.rawValue) {
```

```
        // here we can use the analyzed score
    } else {
        print("Unable to analyze sentiment")
    }
```

`NLTagger` is the primary class we use to process texts in NLP. When we initialize it, we pass the information we are interested in. In our example, we passed `sentimentScore` – a scheme that helps us determine the text sentiment.

Our next step is to set the text input and call the tag function while passing relevant parameters, such as range, unit type, and scheme we want it to analyze.

The tag function performs the text analysis and returns a score between -1 and 1, where a negative score indicates a negative sentiment and a positive score indicates a positive sentiment.

If we run this code on our example sentence before the code example, we'll get a score of 1.0 – an extremely positive text!

Even though text classification is very easy to use, it is also very powerful. We can use this capability to analyze user feedback/reviews, chatbots, and surveys and even adapt an interface, based on the user's sentiments and emotions.

We mentioned that text classification is all about understanding the text sentiment. However, we can use NLP to analyze text using word tagging.

Word tagging

Word tagging is the process of breaking a text into components and assigning tags to each phrase in the text, indicating its grammatical category.

Let's take the example of the following text:

```
She enjoys reading books in the library
```

If we try to break this sentence into grammatical categories, it will be something like *She* (pronoun), *enjoys* (verb), *reading* (verb), *books* (noun), *in* (preposition), *the* (determiner), and *library* (noun).

The different parts of the text are called **tokens**, and their grammatical category is called a **tag**.

The `NaturalLanguage` framework helps us perform tokenization and tag its tokens.

Let's look at the following code:

```
let inputText = "She enjoys reading books in the library"
let tagger = NLTagger(tagSchemes: [.lexicalClass])
tagger.string = inputText
let options: NLTagger.Options =   [.omitPunctuation,
    .omitWhitespace]
```

```
tagger.enumerateTags(in:
  inputText.startIndex..<inputText.endIndex, unit: .word,
  scheme: .lexicalClass, options: options) { tag,
  tokenRange in
    if tag == .verb {
      verb = String(inputText[tokenRange])
      return false
    }
  return true
}
```

The preceding code example takes the same sentence as earlier, tokenizes it, and locates the first verb it finds.

We start by initializing `NLTagger`, similar to what we did in text classification. However, we do that this time by passing `lexicalClass` as its scheme.

Then, we provide the input text and omit punctuation and whitespaces. We do this because we want our text to be as clean as possible. `NLTagger` can catch extra whitespace characters and punctuation as additional tags.

After we clean our text, we call the `enumerateTags` function. This function iterates the words in the text within a given range and extracts the different tags. We compare the tag type inside the passed closure and store it in an instance variable.

In our example, we locate the first verb, which is `enjoys`.

Although word tagging and text classification are `NLTagger`'s two primary use cases, they can also be used for additional cases, such as to identify a text's language:

```
let tagger = NLTagger(tagSchemes: [.language])
tagger.string = inputText
if let language = tagger.dominantLanguage {
    identifiedLanguage =
      Locale.current.localizedString(forLanguageCode:
      language.rawValue) ?? "Unknown"
    } else {
        identifiedLanguage = "Unknown"
    }
```

In the preceding example, the `NLTagger` receives input text and extracts its language. It can identify 50 different languages – impressive for an on-device NLP model!

We can use language identification to identify the user locale and offer to change an app's preferred language, or we can send that information as analytics data to our servers.

Another great example of NLP is **word embedding**. This feature can help our application become smarter.

Each word in the dictionary is related to other words. For example, *house* is related to *building* and *apartment*, and *cat* is associated with *dog*.

We can easily find related words, using a class called `NLEmbedding`:

```
guard let embedding = NLEmbedding.wordEmbedding(for:
   .english) else {
         neighborsText = "Failed to load word
            embedding."
         return
      }
let neighbors = embedding.neighbors(for:
   embedding.vector(for: inputWord) ?? [], maximumCount: 5)
if neighbors.isEmpty {
      neighborsText = "No neighbors found for
         '\(inputWord)'."
      } else {
         neighborsText = neighbors.map { "\($0.0)
            (\($0.1))" }.joined(separator: ", ")
      }
```

In the preceding example, `NLEmbedding` receives an input test, calculates its vector, and finds its closed neighbors. If you ask yourself why this is practical, think of a search engine that can find related content even if it isn't exactly what the user searched for.

In this section, we analyzed text using the `NaturalLanguage` framework. We've learned how NLP works, how to classify text, and extract additional information such as word tagging and even word embedding. However, iOS apps contain more than just text; they also include images. Can we analyze images as well?

Analyzing images using the Vision framework

Analyzing images is a fundamental topic in iOS apps. There are many use cases for analyzing images, such as detecting barcodes, scanning documents, or image editing.

To analyze images in iOS, we need to use Apple's Vision framework. Introduced in 2017 with the release of iOS 11, the Vision framework provides high-level functionality to perform various image analysis tasks.

Understanding how image analysis works

In a way, image analysis works similarly to text analysis, working with different steps that clean and prepare data before inserting it into a model.

The image analysis works with a **CNN (Convolutional Neural Network)**, a neural network designed for visual data.

Consider CNN as a series of filters that can help a model better understand an image. CNN will perform a similar process if the `NaturalLanguage` model preprocessed the text, removing whitespace and duplicate words.

First, the CNN scans an image to detect similar patterns, such as lines, edges, and textures. It then filters out what it thinks are non-important features and shrinks the image to contain the most essential information.

Now that we have a smaller and cleaner image, the CNN tries to decide what the image is – for example, "*It's a cat.*"

Detecting patterns and edges, filtering them, and analyzing an image are complex techniques that require extensive training. Luckily, the Vision framework performs all the heavy lifting for us.

Let's see what it can do for us.

Exploring the Vision Framework's capabilities

Since starting iOS 18, the **Vision framework API** has become extremely simple yet even more powerful.

To understand how the Vision framework API works, we need to remember that it is based on two types – request and observation.

To perform an image analysis, we first create a **request**. Then, we request the specific image and receive an **observation** containing the result (if we have any).

Let's take two popular use cases – detecting barcodes and faces.

Detecting barcodes

Look at the following code to see barcode detecting in action:

```swift
func analyze(url: URL) async {
    let request = DetectBarcodesRequest()
    do {
        let barcodeObservations = try await
            request.perform(on: url)
        barcodeIdentifier =
            barcodeObservations.first?.payloadString ?? ""
    } catch let error {
        print("error analyzing image -
            \(error.localizedDescription)")
    }
}
```

The preceding code block performs barcode detection using the Vision framework. First, we create `DetectBarcodesRequest`, which represents a request to scan barcodes in a given image URL.

Then, we call the request's `perform` function, which returns an array of observations in the case of several barcodes.

Next, we take the first observation payload and store it in a variable. That payload represents the barcode identifier.

Note that the scanning operation can be a heavy task, which is why it is an asynchronous function.

Another interesting example of a Vision framework usage is detecting faces in an image – let's see an example.

Detecting faces

Detecting faces works similarly to detecting barcodes. Let's see a code example:

```
func analyze(url: URL) async {
    let request = DetectFaceRectanglesRequest()

    do {
        let observations = try await
           request.perform(on: url)
        if let observation = observations.first {
            rect = observation.boundingBox.cgRect
        }
    } catch let error {
        print(error.localizedDescription)
    }
}
```

The preceding code example looks almost identical to the previous barcode example. First, we create the request. However, this time, the request is from type `DetectFaceRectanglesRequest`. Next, we perform the detection operation on the given image URL and retrieve an array of observations. Each observation instance contains a rectangle of one of the faces in the image. If the image contains multiple faces, we'll get one observation for each face.

Face detection and barcodes are two common examples of Vision framework use cases. However, the Vision framework is full of surprises and detection capabilities. Let's see what else we can do with it.

Exploring more detection capabilities

As mentioned, the Vision framework is full of machine-learning models capable of detecting almost anything we want. Barcodes and faces are just the tip of the iceberg.

Here's a list of additional detectors:

- **Image aesthetics analysis**: For analyzing an image from an aesthetic viewpoint
- **Saliency analysis**: For finding the most important object in an image
- **Object tracking**: For tracking an object's movement across a sequence of images
- **Body detection**: Similar to face detection, for locating arms, humans, eyes, a mouth, and a nose in images
- **Body and hand pose**: For locating arms in an image as well as detecting their pose.
- **Text detection**: For detecting text in an image
- **Animal detection**: For detecting cats and dogs in an image as well as their pose
- **Background removal and object extraction**: For removing the background and extracting objects from images

The list of the different request types looks impressive, which it is. Reviewing the requests reflects how powerful the Vision framework has become. We can see capabilities usually reserved for high-end image editing applications, such as background removal or object extraction, now available with just three lines of code.

This opens up new possibilities for unique features in our apps, such as working with a camera or prioritizing images based on their information.

We've discussed analyzing text and images, which are considered the most common data sources we usually use. The text and image analysis techniques are different but straightforward to implement.

Now, let's turn to a different type of source we can analyze – sound.

Classifying audio using the Sound Analysis framework

Working with audio is not a popular expertise for many developers. In fact, audio is considered to be a complex and unique world compared to what we developers are used to.

To mitigate this, the iOS SDK also includes an analysis framework that can classify audio using ML models.

Working with the **Sound Analysis framework** differs from the simplicity we are used to with the Vision framework. But don't worry – it is still simple to use.

The Sound Analysis framework contains three different components:

- **SNAudioFileAnalyzer**: The main class that coordinates the analysis work
- **SNClassifySoundRequest**: The sound detection request
- **SNResultsObserving**: A protocol we need to implement to observe the results from the analyzer

To see these three components in action, take a look at the following code:

```
func analyze(at url: URL) {
    do {
        let audioFileAnalyzer = try
          SNAudioFileAnalyzer(url: url)
        let request = try
          SNClassifySoundRequest(classifierIdentifier:
          .version1)

        let resultsObserver =
          ClassificationResultsObserver()
        try audioFileAnalyzer.add(request,
          withObserver: resultsObserver)

        audioFileAnalyzer.analyze()
    } catch {
        print("Error: \(error.localizedDescription)")
    }
}
```

In this example, we first create the `SNAudioFileAnalyzer` instance and initialize it with a URL to the audio file. Then, we create a request for a classification sound request, passing `version1` as a parameter. The `version1` parameter specifies the pre-trained classification version of the model. At the time of writing, no additional versions are available.

Then, we create the `resultsObserver` instance (which we'll discuss briefly) and coordinate everything together, using the analyzer we created earlier.

How do we get the results? Unlike the Vision framework, receiving the results can be streamlined. The `ClassificationResultsObserver` is a custom class that conforms to `SNResultsObserving`. Let's look at the class implementation:

```
class ClassificationResultsObserver: NSObject,
  SNResultsObserving {
    func request(_ request: SNRequest, didProduce result:
      SNResult) {
        guard let result = result as?
          SNClassificationResult else { return }

        if let classification =
          result.classifications.first {
            let result = classification.identifier
        }
    }
}
```

```
        func request(_ request: SNRequest, didFailWithError
           error: Error) { }

        func requestDidComplete(_ request: SNRequest) {}
}
```

The `SNResultsObserving` protocol has three essential request methods – `didProduce`, `didFailWithError`, and `requestDidComplete`.

Great! However, unfortunately, in this case, it seems like we need to go back in time and use the delegate pattern to observe results from the Sound Analysis framework.

The result is a string describing the sound we passed to the analyzer. The code example in this book's GitHub repository shows a sound file with a baby crying. In this case, the result would be `baby_crying`.

Apple has yet to officially publish the number of sound classes that the Sound Analysis framework can recognize. However, in most cases, this should be enough for day-to-day usage.

The Sound Analysis framework can be great for monitoring apps, adding **SDH (subtitles for the deaf or hard of hearing)** to video captions, and analyzing videos.

So far, we have discussed how to analyze different types of data – sound, images, and text. However, ML is valuable in other areas, such as app search.

Performing a semantic search with Core Spotlight

When we discussed NLP in the *Interpreting text using NLP* section, we said that one of the most common NLP use cases is analyzing a search phrase to build intelligent search queries.

Even though the `NaturalLanguage` Framework API is robust and straightforward, performing a semantic search is considered a complex task.

Starting iOS 18, the Core Spotlight framework supports a semantic search. Before we dive into the details, let's clarify the term **semantic search**.

Understanding what semantic search is

Let's think together about how search queries work in a standard app, and we'll do that using an example.

Imagine that we have a course catalog app where a user can search for a particular course, and let's say we have the following list of courses in our local data store:

- Management for employees
- Data science
- Digital marketing
- ML and AI

Our user wants to improve their leadership skills, so they search for a management course within this list of courses.

The search query's basic form is to match a specific phrase. For example, if the user searches for `management`, we filter only courses containing *management*. We also need to ensure that the output query is case-insensitive.

However, what if the user searches for `manager`? In this case, our query returns no results, even though a typical user can search for *manager* if they want a course about management.

In this case, we can use the `NaturalLanguage` framework to try and perform lemmatization of the search phrase. **Lemmatization** is a technique that reduces words to their basic form. So, the basic form of *manager* is *manage*.

However, if we want to match the search phrase *manage*, we also need all our records with the word *management* to contain the word *manage* so that we can filter the results accordingly. It means we must maintain the basic form for each word in each record.

But things can get even more complex than that. What if the user searches for a management course using the phrase *leadership*? In this case, we will have to index our records with embedded words, as we learned in the *Word tagging* section of this chapter.

The conclusion is that basic search is easy. However, semantic search, which is much more effective, is also much more complex.

As mentioned, semantic search is built on top of the **Core Spotlight framework**, starting with iOS 18. The Core Spotlight framework is not new – it was introduced in 2015 as part of iOS 9 and helps developers index app content and make it searchable, using the Spotlight feature in iOS.

This chapter does not cover using the Core Spotlight framework. However, we will briefly review the Core Spotlight principles to understand how to enable semantic search. Let's begin.

Exploring the Core Spotlight framework

The Spotlight framework indexes local data and retrieves it by performing queries.

The Core Spotlight framework has three primary parts – creating searchable items, indexing, and querying. Let's go through the parts one by one.

Creating searchable items

Let's say we have instances of a book structure in our local storage and want to implement Core Spotlight to allow users to search for books.

First, we need to map all our `Book` instances to `CSSearchableItem`:

```
let searchableItems: [CSSearchableItem] = books.map { book
  in
        let attributeSet =
          CSSearchableItemAttributeSet(contentType:
          .text)
        attributeSet.title = book.title
        attributeSet.contentDescription = book.author

        let item = CSSearchableItem(uniqueIdentifier:
          book.id, domainIdentifier: "books",
          attributeSet: attributeSet)
        return item
}
```

In the preceding code example, we took an array of `Book` and mapped it to an array of `CSSearchableItem`. We do that by creating a `CSSearchableItemAttributeSet` – an item that contains general information about the searchable item. Then, we initialize a new `CSSearchableItem`, passing our `CSSearchableItemAttributeSet` and providing a unique identifier that can help us retrieve the `Book` record when needed.

Indexing

Now that we have an array of `CSSearchableItem`, we need to index the array items for the Core Spotlight framework. We do that by creating `CSSearchableIndex`:

```
let index = CSSearchableIndex(name: "SpotlightSearchIndex")
      index.indexSearchableItems(searchableItems) { error
        in
          if let error = error {
              print("Indexing error:
                \(error.localizedDescription)")
          } else {
              print("Books successfully indexed!")
          }
      }
```

In the preceding example, we created a new `CSSearchableIndex` and called the `indexSearchableItems` function, with the array of `CSSearchableItem` that we made in the previous step. Note that this is an asynchronous operation and is considered to be quite intensive.

Querying

Now that we have an index, we can perform a query to retrieve data based on a search phrase:

```
let searchContext = CSUserQueryContext()
        searchContext.fetchAttributes = ["title"]
        searchContext.enableRankedResults = true

        var items: [CSSearchableItem] = []
        let query = CSUserQuery(userQueryString: query,
          userQueryContext: searchContext)
        do {
            for try await element in query.responses {
                switch(element) {
                case .item(let item):
                    items.append(item.item)
                    break
                case .suggestion(let suggestion):
                    // handle suggestions.
                    break
                @unknown default:
                    break
                }
            }
            self.searchResults = items
        } catch let error {
            print(error.localizedDescription)
        }
```

In the preceding example, we create a search context containing various query information. Based on the search context and the search phrase, we initialize an item of `CSUserQuery` and fetch the search results by calling its `responses` getter.

The results are an array of `CSSearchableItem`, and we can retrieve the original item by using the unique identifier for each record.

Now that we know how to implement search using the Core Spotlight framework, let's see how to implement a semantic search.

Implementing semantic search

Adding semantic search capabilities to an existing Core Spotlight search is simple. All we need to do is load the ML model once using the following static function:

```
CSUserQuery.prepare()
```

The `prepare` function prepares the Core Spotlight framework to load its ML models for semantic search.

If the search index has a protection level due to privacy concerns, we also need to call the `prepreProtectionClasses` function:

```
CSUserQuery.prepareProtectionClasses([.completeUnlessOpen])
```

This function prepares the search for indexes marked with the `completeUnlessOpen` protection level.

> **What are protection levels?**
>
> The term **protection level** refers to the accessibility level where users have specific resources, considering the device's security conditions. There are three primary protection levels:
>
> - `NSFileProtectionNone`: The index is always accessible, even when the device is locked
>
> - `NSFileProtectionCompleteUntilFirstUserAuthentication`: Once the user is authenticated for the first time after a device restart, the index is accessible
>
> - `NSFileProtectionComplete`: The index is accessible only when the device is unlocked

Remember that preparing the ML models costs time and memory, so it's better to call the `prepare` function only immediately before the search user interface.

We have discussed various built-in ML models, and we can see that they cover many use cases where we can use ML capabilities with our projects. However, there are cases where the iOS SDK doesn't provide the exact ML solution we need. Luckily, we can integrate our models using the CoreML framework.

Integrating custom models using CoreML

Generally, ML models are trained to perform a specific task – recognizing a sentence's sentiment, detecting humans, or analyzing sounds are all examples of different tasks done using various models. This means that even though the potential of the existing models is enormous, we are still limited in what we can do.

This is where the **CoreML framework** enters the picture. Using CoreML, we can integrate ML models that are not part of the iOS SDK, and we can even train our own models and add more intelligent capabilities.

It's best to explain how to do this by using an example, such as detecting spam messages.

Imaging we are developing a messaging app. One of the most popular messaging app features is the ability to detect spam to improve the user experience and increase retention.

We must create an ML model to classify messages as spam to implement a spam detector.

To achieve this, we can use a desktop application called Create ML, which is part of the Xcode suite. Let's begin by learning more about Create ML!

Getting to know the Create ML application

Create ML was introduced in 2018 as part of Apple's ongoing effort to make ML more accessible to developers. We can build, train, and deploy ML models in various areas using Create ML.

To open Create ML, follow these steps:

1. Open Xcode.
2. Right-click on the Xcode icon on the dock.
3. Select **Open Developer Tool | Create ML**.

Another way to open `Create ML` is by searching for it in Spotlight on your Mac and selecting it.

After opening it and clicking on the **New Document** button, we get the following screen (*Figure 12.1*):

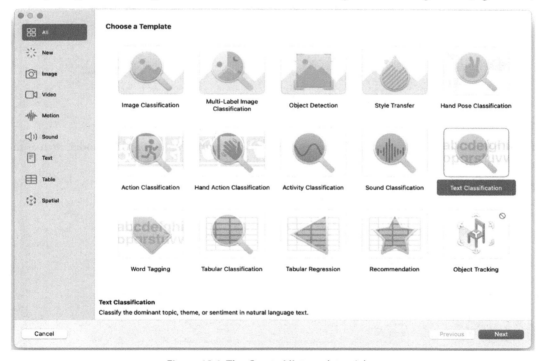

Figure 12.1: The Create ML template picker

Figure 12.1 shows the Create ML template picker screen. Each template represents a different configuration for our model, and each is designed to handle a different type of data. For example, the **Image Classification** template is designed to handle images. Since we want to classify text messages, we will pick the **Text Classification** template and click the **Next** button.

This will take us to the project details screen (*Figure 12.2*):

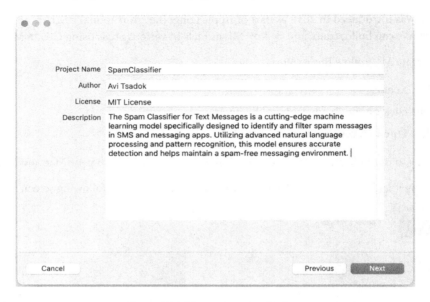

Figure 12.2: The project details form

In the project details form, we will fill in some general information about our ML model, such as the name, author, license, and description, and then click **Next**.

Our next screen is the project window (*Figure 12.3*):

Figure 12.3: The SpamClassifier project window

In *Figure 12.3*, we can see the `SpamClassifier` project window. The project window is the main window where we will build our model. Let's go over the different window components:

- **Left pane**: The left pane lists the project's different sources – the ML model and its data sources, used for training and testing
- **Settings tab**: The **Settings** tab is where we define the different data sources for the various phases and general training parameters
- **Training tab**: The **Training** tab shows the progress of the training operation
- **Evaluation tab**: The **Evaluation** tab shows the performance of our model in the different phases
- **Preview tab**: We can *play* with our ML model and experience it in the **Preview** tab
- **Output tab**: The **Output** tab is the place where we can deploy our model

The list of the components reflects the phases we must go through when we build our model.

Now that we know what Create ML is, let's start building our model.

Building our Spam Classifier model

Our Spam Classifier model-building process is based on three data sources – training, validation, and testing data. These three data sources are something we covered earlier in this chapter in the *Training the model* section.

First, let's take a look at how we will prepare our data.

Preparing our data

Since we are building a Spam Classifier model, we must prepare a dataset containing both spam and non-spam messages. The text classification template requires our dataset to be in the form of a CSV file with two columns – `text` and `label`. In our case, the `text` column represents the content of the SMS message, and the `label` column is the classification – `true` for a spam message and `false` for non-spam.

The ratio between the spam and the non-spam messages needs to reflect the real-world distribution. In our case, we have a dataset file with 300,000 records, where 10% of them are spam messages and 90% are non-spam messages.

To set the **training dataset**, we can drag the CSV file into the **Training Data** box (*Figure 12.4*):

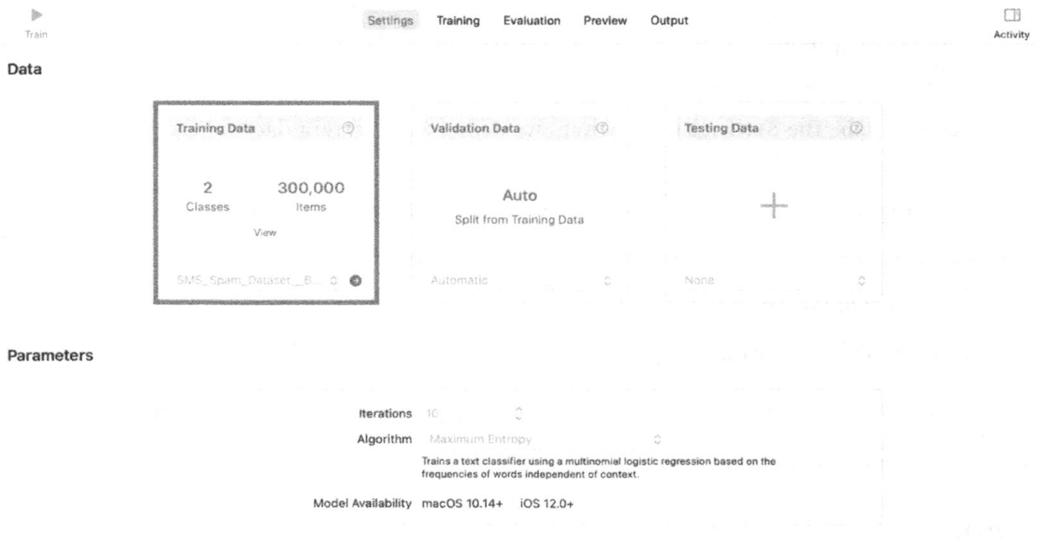

Figure 12.4: Training Data with 300,000 records

Figure 12.4 shows the **Settings** tab, with the training data now containing 300,000 records with 2 classes. The classes are `true` and `false`, as stated earlier. In addition, we also have a new data source in the left pane – the file we imported as the training dataset.

We can handle the **validation data** now that we have the training data. As a reminder, as part of the training process, we will use the validation data to tune the model. We can provide our own validation data, but Create ML allows us to split it from the training dataset we've just supplied.

The third dataset is the **testing data**. We use the testing data to see how the model classifies unseen text. We can add the testing dataset later in the evaluation step.

Apart from choosing the different datasets, we can also set the number of iterations our training will go through and the model algorithm.

With each iteration, the training process can tune itself by reviewing the errors from the previous iteration and adjusting its parameters (like weights in a neural network). Our intuition may say that the more iterations we have, the more our model will be smarter. However, this is not so. First, at some point, having another iteration stops improving the model and only consumes computational resources. But the real problem is what we call overfitting. **Overfitting** is when an ML model learns the training data too well, including its noise. In this case, there will be issues with analyzing unseen data.

Another parameter is the model algorithm (*Figure 12.5*):

Figure 12.5: Choosing the model algorithm

Figure 12.5 shows the pop-up menu where we can choose the model learning algorithm from five different options. The algorithm overview is not in this chapter's scope, but in short, different algorithms are suitable for different needs and consume other resources. For example, the **BERT** algorithm is ideal for semantic understanding, and the **Conditional Random Field** is great for sequence labeling. In our case, we will choose the **Maximum Entropy** algorithm, which is excellent for classification.

Now that we have all our datasets ready, we can click the **Train** button in the top-left corner and start our training.

Performing the training

Now, we have arrived at the main dish – the training phase. In the training phase, the Create ML app goes over the training dataset using the algorithm we defined in the *Preparing our data* section. Let's try to describe that process:

- In each iteration, the model *verifies itself* using the validation dataset. Remember that the validation dataset can be distinct. However, by default, it is a subset of the training dataset.
- The *duration* of the training phase is derived from three major factors – the dataset size, the number of iterations, and the chosen algorithm.
- The model doesn't have to perform the number of iterations we defined in the **Settings** tab. If the validation accuracy reaches a high level, *the training will stop earlier* to save resources and avoid overfitting.

At the end of the training process, we'll see the following graph (*Figure 12.6*):

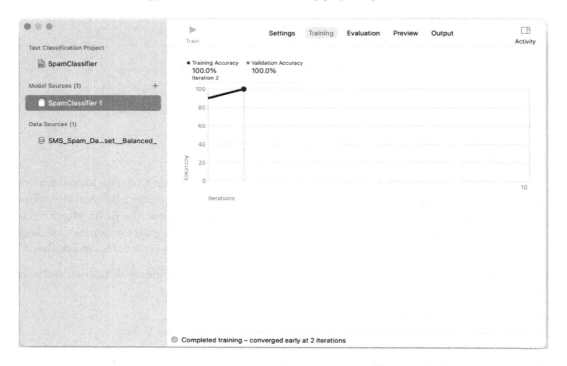

Figure 12.6: The Training tab at the end of the training process

Figure 12.6 shows how well we did in our training phase. We can see that we have reached a high accuracy after only two iterations. In this case, it is because our training dataset is well-structured and reliable. However, that won't always be the case, so we need patience in this step.

Now that our model has been trained, we need to test it. To do that, we will use our test dataset as part of the evaluation step (*Figure 12.7*):

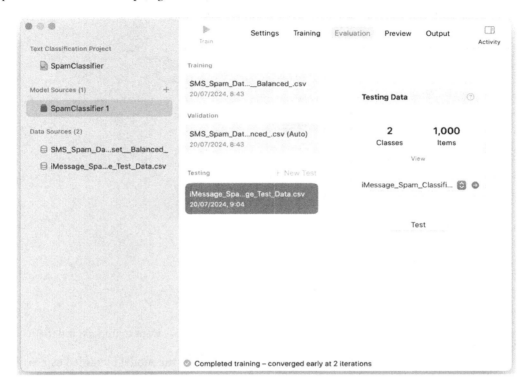

Figure 12.7: The model evaluation step

Figure 12.7 shows the evaluation step and the different datasets used to train and validate the model. We can also see that the testing data contains a dataset of 1,000 items. The testing dataset structure is similar to the training and validation datasets. Tapping on the **Test** button runs the classification on all the 1,000 items in the dataset and measures their classification accuracy. Let's see the test result (*Figure 12.8*):

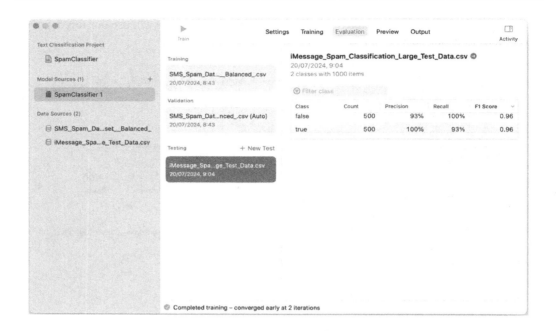

Figure 12.8: The evaluation results

Figure 12.8 presents some terms that we need to be familiar with if we want to understand the report:

- **Precision**: Precision is the percentage of all messages that the model identified as `true` or `false` (depending on the specific class) and that were correct. For example, 93% precision for the `false` class means that 93% of all the messages the model identifies as `false` were actually `false`.

- **Recall**: Recall is the counterpart to precision. A recall of 93% for the `true` class means that the model correctly identified 93% of all actual spam messages.

- **F1 Score**: The F1 score is the balance between precision and recall.

The **F1 Score** involves more than just measuring a model's accuracy. It balances two important metrics – **precision** and **recall** – and reflects a better model performance measurement. In our case, a score of 0.96 is considered a very high performance.

Our next tab is **Preview**, where we can play within a playground zone (*Figure 12.9*):

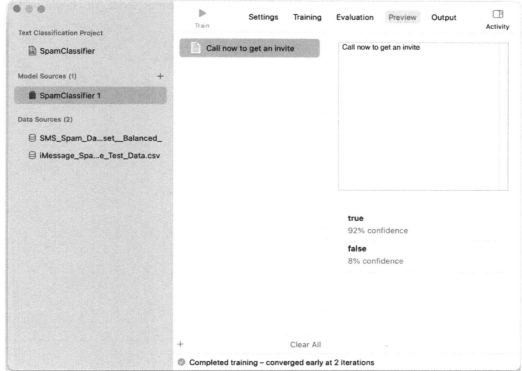

Figure 12.9: The Preview tab

Figure 12.9 shows our model's **Preview** tab, with an example message that says, **Call now to get an invite**. Our model identified this message as spam with a 92% confidence. Good job!

Now, let's see how we can deploy our model.

Deploying our model

There's no point in having a great training process if we can't deploy it in Xcode. This is why we have the **Output** tab (*Figure 12.10*):

Figure 12.10: The Create ML Output tab

The **Output** tab shows a summary of our model, including a new detail we haven't seen until now – the model size.

More importantly, the **Output** tab also contains the option to export a model or open it in Xcode. Clicking the **Get** button allows us to save the model locally in a file with an `mlmodel` extension.

To use the `mlmodel` extension in our projects, we'll need to use Core ML. That's our next topic.

Using our model with Core ML

The **Core ML framework**'s goal is to allow us to integrate ML models into our projects.

Our first step is to add the `mlmodel` file that we saved from the Create ML application to Xcode. We can do that by dragging the file to the project navigator in Xcode.

The main class in the Core ML framework we will use is `MLModel`, which represents a ML model loaded into the system. To load our Spam Classifier model, we initialize the model in our code:

```
class MessageClassifier {
    let model: MLModel

    init(configuration: MLModelConfiguration =
      MLModelConfiguration()) throws {
        model = try SpamClassifier(configuration:
          configuration).model
    }
}
```

In the preceding code example, we created a new class, called `MessageClassifier`, which encapsulates our ML integration with the Spam Classifier model.

We then initiate the class, passing a new `MLModelConfiguration`. This contains different options, but we can pass an empty instance at this stage.

Our class also contains an `MLModel` instance. To initiate the model instance, we use the `SpamClassifier` class, passing our configuration.

But wait – where did the `SpamClassifier` class come from?

When we added the Spam Classifier `mlmodel` file into our Xcode project, Core ML generated three interfaces – the `SpamClassifier` class, `SpamClassifierInput`, and `SpamClassifierOutput`.

Now that we have our model, let's write a function that can predict whether a message is spam:

```
func prediction(text: String) throws -> Bool {
        let input = SpamClassifierInput(text: text)
        if let result =  try? model.prediction(from: input)
        {
            let value = result.featureValue(for:
              "label")!.stringValue
            return value == "true"
        }
        return false
    }
```

In the preceding example, we created a `prediction` function that receives a text message as input and returns a Boolean.

It starts by creating a `SpamClassifierInput` instance with the text input. Then, it generates a prediction result for this input by running the model's `prediction()` function. We then get the value from the feature, called `label`, and compare it to `true`.

This code example demonstrates how to easily use a custom ML model in our Xcode projects.

Now, let's try to understand if using a custom ML in our Xcode is that simple.

Where to go from here

The Core ML part of this book is unique. In most cases, I have simplified complex topics to make them more accessible for developers. However, I think the Core ML topic is different.

ML is a broad topic, beyond the scope of this chapter. Furthermore, it is a complex topic. Training is more than just delivering datasets. It is essential to understand the dataset mix between the different classes, pick the correct algorithm, and read the evaluation results carefully.

And remember that the model we created is a custom. This means that we don't have any control over how its algorithm works and need to observe and fine-tune it over time.

In summary, ML is a complex topic, and if we want to enter this area, we need to approach it more in-depth than reading 15 pages.

Summary

This was a long but fascinating chapter about one of the most exciting contemporary topics – ML and AI.

We reviewed the basics of AI and ML, understanding what it means to train an ML model. We explored the built-in ML models in the iOS SDK, including NLP, analyzing images using the Vision framework, and classifying audio with the Sound Analysis framework. We learned how to add semantic search capabilities to the Core Spotlight framework, and if that wasn't enough, we also learned how to create and integrate custom ML models into our projects.

Now, we can add some intelligence features to our apps!

Speaking of intelligence, our next chapter discusses how we can integrate Siri using App Intents. The ML phase is not over just yet!

13
Exposing Your App to Siri with App Intents

For many years, apps have lived and operated alone in system space. Each app is totally isolated from the others, without the capability to communicate or expose data.

Over the years, things have changed a bit. One of the most exciting features apps gained was enhancements in **App Intents**. At this point, you should be familiar with App Intents—we discussed them in *Chapter 5*. However, in iOS 18, App Intents became even more powerful as they worked closely with Apple Intelligence and not just with WidgetKit. That's why we are going to cover App Intents in more detail.

In this chapter, we will learn about the following:

- Understanding the App Intents concept
- Creating a simple app intent
- Formalizing our content using app entities
- Adjusting our app intents to work with Apple Intelligence

The ability of App Intents to open up our app is truly remarkable and full of potential. But first, let's understand the App Intents concept.

Technical requirements

For this chapter, it's essential to download Xcode version 16.0 or higher from the App Store.

Ensure you're operating on the most recent version of macOS (Ventura or newer). Just search for Xcode in the App Store, choose the latest version, and proceed with the download. Open Xcode and complete any further setup instructions that appear. After Xcode is completely up and running, you can begin.

This chapter includes many code examples, some of which can be found in the following GitHub repository: https://github.com/PacktPublishing/Mastering-iOS-18-Development/tree/main/Chapter%2013.

Notice that some examples in this chapter require running on a modern device, such as an iPhone 15 Pro/Max, iPad with M1 and above, or Apple Silicon Mac.

Understanding the App Intents concept

We first encountered App Intents in *Chapter 1* and then in *Chapter 5*, when we discussed WidgetKit. But do we really understand the concept of App Intents? Let's get a short background about Apple's efforts to integrate AI deeply inside the system across many apps, including third-party apps.

To implement this integration, we need to create an API for our app that exposes the app's core content and main actions. For example, a to-do app can create an API that lets Siri or other system components create a new task, complete an existing task, or pull the list of tasks stored in core data. A delivery app can have an API that returns an answer to whether the delivery services are now open or at what time the delivery arrives.

This API, called App Intents, is our way of exposing our app's main use cases and content to the world. It's one of the tools we use to integrate our app with Apple Intelligence as well.

If this sounds complex, you'll be surprised how simple it is to create an app intent. Let's see how it works.

Creating a simple app intent

To demonstrate how to create an app intent, let's imagine we have an amazing to-do list. Not just amazing, even mighty! So, we'll call it `MightyTasksList`.

Our `MightyTasksList` app is so great that our users demand that they use it with Siri while they are driving. So, we decided to create an app intent.

To do that, we'll open a new file and write the following code:

```
import AppIntents

struct GetTasksIntent: AppIntent {

    static var title: LocalizedStringResource { "Get the number of opened tasks" }

    @MainActor
    func perform() async throws -> some ProvidesDialog {
        let tasks = TaskManager().tasks
        return .result( dialog: "Number of the opened tasks is \(tasks.count)")
```

        }
}

This is it? Yes! Writing a simple app intent is extremely easy. Let's narrow down what we did here:

- We imported the `AppIntents` framework. In this case, we need this framework to have the `AppIntent` protocol.
- We created a `GetTasksIntent` structure that conformed to the `AppIntent` protocol. This structure defines our intent functionality.
- As part of the `AppIntent` protocol, we must implement two things. The first one is the intent's `title`, which appears in the intent gallery in the Shortcuts app (we'll get there shortly in the *Running the intent in the Shortcuts app* section). The second thing we need to implement is the `perform()` function, which is the actual code that will be executed when the intent runs.

The `perform()` function always returns an `IntentResult` protocol-based type—in this case, we return a `ProvidesDialog` instance that conforms to the `IntentResult` protocol and displays a message to the user. However, there are additional types, and we'll discuss them in the following sections.

Next, let's run our intent using the Shortcuts app.

Running the intent with the Shortcuts app

The **Shortcuts app** is a powerful automation tool that allows users to create shortcuts for routines and actions in their system. Users can also use the Shortcuts app to create scripting, automation, conditional statements, and complex logic.

Apple acquired the Shortcuts app back in 2017, which was initially developed by a start-up called Workflows.

When we build and run our app, iOS scans for structures that conform to the `AppIntent` protocol and add them to the **Shortcuts** gallery in the Shortcuts app (*Figure 13.1*):

Figure 13.1: Our intent is shown in the Shortcuts app

In *Figure 13.1*, we can see the intent of the **Get number of the opened tasks** shortcut in the Shortcuts app when searching for action for our app. Then, we can add the action as a new shortcut. Let's see what happens when we run our intent (*Figure 13.2*):

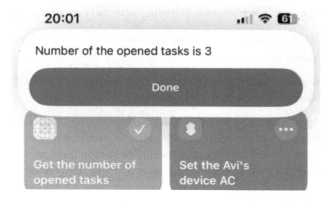

Figure 13.2: Running the Get number of opened tasks shortcut

Figure 13.2 shows what happens when we run our app intent. We can see the message we defined as part of the `perform()` function results.

Way to go! We created our first app intent!

Now, let's try to ease our users and create the shortcut as part of the app.

Creating an app shortcut

Instead of letting the user create a shortcut based on the action (intent) we provide, we can create a shortcut for them to use.

To do that, we need to create a structure that conforms to the `AppShortcutsProvider` protocol to create a pre-configured shortcut:

```
import AppIntents

struct AppShortcuts: AppShortcutsProvider {
    @AppShortcutsBuilder
    static var appShortcuts: [AppShortcut] {
        AppShortcut(intent: GetTasksIntent(),
            phrases: ["What is left in \(.applicationName)?", "How many tasks left in \(.applicationName)"], shortTitle: "My tasks",
systemImageName: "circle.badge.checkmark")
    }
}
```

In this code, we have a struct called `AppShortcuts`. This struct has one variable to implement—`appShortcuts`, which contains a list of the app shortcuts.

In this case, we create a new `AppShortcut` instance that contains the following:

- The intent that will be executed. Here, we put the intent we created in the preceding section (`GetTasksIntent`).
- The exact phrases the user has to say to Siri. In our case, we added two phrases. Notice that the phrases must include the application name.
- A title and a system image for the shortcut.

Once we run our app, the user doesn't need to create a shortcut using the Shortcuts app—the shortcut is ready for the user to use with Siri.

Now that we have created our first intent and shortcut, let's dive in for more complex use cases.

Adding a parameter to our app intent

In the *Creating a simple app intent* section, we created a `GetTasksIntent` struct that went to the persistent store and returned the user's number of open tasks. Now, let's see how we can use the `AppIntents` framework to create an action that inserts a new task into the system.

We'll open a new file and add the following code:

```
struct AddTaskIntent: AppIntent {
    
    static var title: LocalizedStringResource { "Create new task" }
    
    @Parameter(title: "Title")
    var title: String
    
    @MainActor
    func perform() async throws -> some ReturnsValue<String> {
        TaskManager().addTask(Task(title: title))
        TaskManager().saveTasks()
        return .result(value: title)
    }
}
```

`AddTaskIntent` is a little bit more complex but not too complex.

To start, we create a new `AddTaskIntent` that conforms to the `AppIntent` protocol, similar to `GetTasksIntent`. We also provide a readable title for the Shortcuts app. But then we see a new variable—a title marked with a `@Parameter` attribute.

We said that app intents are actually our application API. Some APIs require input, so when adding a new task, the title is our Intent input. We can provide this input using Siri, a dialog, or even another intent. When we run `AddTaskIntent`, the user must provide a title for the task.

When we reach the `perform()` function, we can see how we use the `title` parameter to insert a new task into the persistent store. We can also see a change compared to the `GetTasksIntent` example in how the function returns a value.

In `GetTasksIntent`, we used the `ProvidesDialog` protocol. Now, we use `ReturnsValue<String>`, which returns a value to our shortcut. We can use the returned value as input for other actions. For example, in this case, we can use the task title to create a reminder in the Reminders app with the same title or even send a message with this title to someone else. This feature makes App Intents and the Shortcuts app extremely useful for power users.

Let's see how it looks in the Shortcuts app:

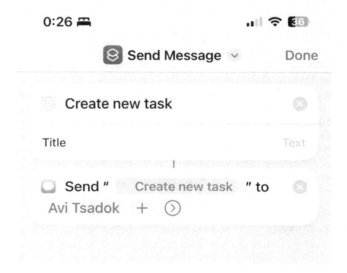

Figure 13.3: Shortcuts with two actions

In *Figure 13.3*, we can see our shortcut with two actions:

1. The first is `AddTaskIntent` from our app.
2. The second is that the result of our intent is the task title and the input of the *send message* action from the messages app.

We can see that it is possible to chain actions together and create powerful streams.

Let's see how it looks when we run our shortcut (*Figure 13.4*):

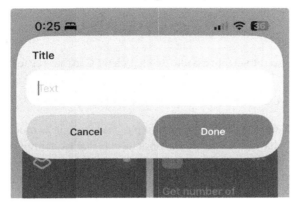

Figure 13.4: Running AddTaskIntent

Figure 13.4 shows that the system asks for the task title when we run our shortcut by providing a standard input field. Besides the Siri integration, this is part of what we get for free—a standard user interface that handles all of that for us.

However, we can also create our own user interface for the shortcut! Let's see how to do that.

Returning a custom view

We have added two important intents in the previous sections. The first receives a string containing the number of opened tasks, and the second is an app intent that creates a new task in the persistent store.

Let's discuss another use case—getting the list of opened tasks. In this case, we want to present the user with a custom view since Shortcuts and Siri don't know how to deal with a list of entities natively.

So, let's create a custom view:

```
struct MiniTasksList: View {
    let tasks: [Task]

    var body: some View {
        VStack {
            ForEach(tasks) { task in
                TaskView(task: task)
            }
        }
    }
}
```

Our code example contains a struct named `MiniTasksList` with a VStack that displays an array of `TaskView`.

There are two weird things here:

1. First, why do we need to create a dedicated list view? Can't we reuse the view we already have in our app?
2. Second, why do we use a VStack and not a List view?

These arguments are usually valid. We should aim to reuse our code as much as possible and use the right view for the right behavior. However, one limitation is that we cannot use List or Scroll views as part of our custom view. We also can't display animations or allow user interaction. If we want to achieve more functionality, we should use the app itself or create additional app intents to fulfill our needs.

Now that we have a custom view, let's create an app intent that uses it:

```
struct GetTasksListIntent: AppIntent {

    static var title: LocalizedStringResource { "Get my Tasks's List"
}

    @MainActor
    func perform() async throws -> some ShowsSnippetView {
        let tasks = TaskManager().tasks
        return .result(view: MiniTasksList(tasks: tasks))
    }
}
```

The `GetTasksListIntent` structure is similar to the previous intents we created in the previous sections. It also has a `title` property and a `perform()` function.

There are two important changes here:

- The return type of the `perform()` function is now `ShowsSnippetView`. We use `ShowsSnippetView` if we want to present a custom view as a result of our function.
- We returned a different intent result with the `MiniTasksList` view we created.

Now, let's run our intent using the Shortcuts app and see what happens (*Figure 13.5*):

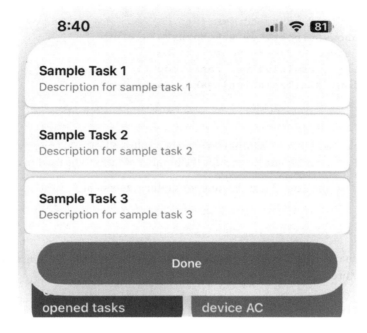

Figure 13.5: The list of tasks as part of the intent response

Figure 13.5 shows `MiniTasksList` as our intent response. Looking at how the list is displayed, we can understand why Apple limits how we can customize this view. The goal is for our view to be as simple as possible and aligned with the rest of the intents.

Returning a view is great. But what if we want to return a custom view and a value that can be used for other purposes? Is it possible? Let's find out if and how to do that.

Having multiple result types

Imagine that, besides getting the tasks list, the user wants to add another step to its shortcut. If the number of tasks is bigger than, let's say, five, they want to open the calendar to re-arrange the day.

So, we would like to show the list of tasks and also return their quantity.

We can do that by returning multiple types:

```
struct GetTasksListIntent: AppIntent {
    static var title: LocalizedStringResource { "Get my Tasks's List"
}
```

```
    @MainActor
    func perform() async throws ->
      some ShowsSnippetView & ReturnsValue<Int> {
        let tasks = TaskManager().tasks
        return .result(value: tasks.count,
          view: MiniTasksList(tasks: tasks))
    }
}
```

In our code example, `perform()` returns two types of results – `ShowsSnippetView` for displaying the tasks list and `ReturnsValue<Int>` for the number of tasks to be used in other intents.

We also changed `IntentResult` at the function's return statement:

```
return .result(value: tasks.count,
    view: MiniTasksList(tasks: tasks))
```

Notice that the value we return needs to be the same as the `ReturnsValue` instance we declare in the function signature.

Adding confirmation and conditions

Having the app perform actions can lead to more complex use cases. For example, there are cases where we need to confirm a specific action with the user, such as deleting or ordering something. In other cases, we might want to inform the user that we cannot perform the action or even request more information.

The `AppIntents` protocol has the capability to create a dialog with our users. This dialog can be used with Siri to make the process feel more conversational.

Let's go back to our to-do app and create an app intent that allows the user to delete all of its tasks:

```
struct DeleteAllTasksIntent: AppIntent {
    static var title: LocalizedStringResource { "Delete all tasks" }

    func perform() async throws -> some ProvidesDialog {

        let taskManager = TaskManager()
        if taskManager.tasks.count == 0 {
            return .result(dialog: .init("Sorry, there are no tasks to delete"))
        }

        try await requestConfirmation(actionName: .go,
```

```
            dialog: IntentDialog("Are you sure you want to delete all
your tasks?"))

        TaskManager().deleteAllTasks()
        return .result(dialog: .init("All of your tasks have been
deleted."))
    }
}
```

This intent is a bit more complex and smarter than our previous examples. At the beginning of the `perform()` function, we check to see if there are any tasks in the persistent store to delete. If there are no tasks to delete, we notify the user by returning a simple text dialog.

Next, since it is a destructive action, we want to confirm it with our user. So, we use the `requestConfirmation()` function. This function presents a dialog with a given text and a confirmation button (*Figure 13.6*):

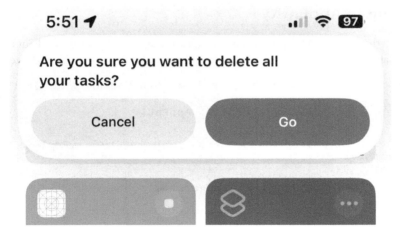

Figure 13.6: A confirmation dialog

Figure 13.6 shows a confirmation dialog that was derived from the `requestConfirmation()` function. Notice that we can choose from a set of confirmation button titles. In our case, we selected the **Go** title.

The next steps are straightforward: We perform the deletion action and notify our user that it has been executed.

Up until now, our app intents returned primitive types such as strings and int. But what about working with our app types? Is it possible to transfer them as part of the shortcut actions chain? This is what `AppEntity` is for.

Formalizing our content using app entities

In our app intents, when we created a task, we returned a string value of the task's title. However, a task is more than just a title – it contains a description, status, ID, and many more properties. In other words, a task is not a string or a **Universally Unique Identifier (UUID)** value type – it's a structure named `Task`.

The problem with App Intents is that no other app or the system knows what a `Task` structure is since it's our app's internal type. We need to expose the type to the system intent world to make `Task` a known type by the system.

Let's connect that to a use case: creating and opening a task in the app. To make it modular, we want to create two intents: creating a task and opening a task. When we have the two intents, we can chain them in a shortcut.

Let's start by letting the system know what `Task` is.

Conforming to AppEntity

Conforming to the `AppEntity` protocol makes app entities available to Siri and Shortcuts. It means that when our app intent returns one of our entities, we can pass it as input to the next intent in the chain.

Let's see how we can take our `Task` structure and make it conform to `AppEntity`:

```
struct Task: Identifiable, Codable, AppEntity {

    static var typeDisplayRepresentation: TypeDisplayRepresentation {
.init(stringLiteral: "Task") }

    init(id: UUID = UUID(), title: String,
      description: String = "") {
        self.id = id
        self.title = title
        self.description = description
    }

    var displayRepresentation: DisplayRepresentation {
DisplayRepresentation(stringLiteral: "title: \(title)") }

    let id: UUID

    @Property(title: "Title")
    var title: String

    @Property(title:"Description")
```

```
    var description: String


    static var defaultQuery = TaskQuery()
}
```

Let's break down the AppEntity protocol implementation:

- typeDisplayRepresentation: Our entity needs to have a name in the system so we can display it in the Shortcuts app. In this case, we return Task.
- displayRepresentation: While typeDisplayRepresentation shows the entity type name, the displayRepresentation property returns the entity value representation. In this case, this is the title value (e.g., *Call my mom*).
- **Exposing the entity's properties using the @Property attribute**: When we add the @Property attribute to some of the entity's properties, we define the entity structure for use in the Shortcuts app.
- defaultQuery: Declaring our app's entities is not enough; we also need to provide the system with a way to retrieve them. Our next step will be to create the query that the system will use to fetch our entities.

Now that our Task struct is known by the system, let's finish the implementation by creating TaskQuery:

```
struct TaskQuery: EntityQuery {
    func entities(for identifiers: [UUID]) async throws -> [Task] {
        return TaskManager().tasks.filter {identifiers.contains($0.
id)}
    }

    func suggestedEntities() async throws -> [Task] {
        return TaskManager().tasks
    }
}
```

In this code example, we can see that the TaskQuery structure conforms to the EntityQuery protocol.

The system uses the first function, entities(), to retrieve the entities by identifiers. At this point, we reach the app services (in this example, TaskManager) to fetch, filter, and return an array of entities. That's why this function is required.

The second function (suggestedEntities()) is not required, but it can help the system present the user with a list of entities while we are fetching the list of entities.

We know how to define AppEntity and its query at this point, but we need to connect it to an app intent to understand how they are being used.

Let's do that by creating an `Open a task` intent.

Creating an Open a task intent

Creating an `Open a task` intent is not that different from what we saw in the previous examples. This time, we'll integrate the new app intent with the `AppEntity` struct we've just created:

```
struct OpenTaskIntent: AppIntent {
    static var title: LocalizedStringResource { "Open a task" }

    @Parameter(title: "Task")
    var task: Task?

    static let openAppWhenRun: Bool = true

    @MainActor
    func perform() async throws -> some ProvidesDialog{
        let taskToOpen: Task

        if let task {
            taskToOpen = task
        } else {
            taskToOpen = try await $task.requestDisambiguation(
                among: TaskManager().tasks,
                dialog: "What task would like to open?")
        }

        Navigator.shared.path.append(taskToOpen)

        return .result(dialog: "Opening your task")
    }
}
```

Our `Open a task` intent is structured like our previous intent examples. Still, there are additional changes we need to discuss:

- We added `@Parameter` to `Task`. Using `@Parameter` is not new to us—we discussed it in the *Adding a parameter to our app intent* section. However, this time, we do that with the `Task` structure itself. We can do that because `Task` now conforms to `AppEntity`.
- We set `openAppWhenRun` property to `true` so we can open the app and display the task details.
- If the app intent doesn't receive a task parameter, we can ask the user to select a task using the `requestDisambiguation` function. This function presents a dialog to the user with a given list of tasks and asks them to select a task.

After we have a task, we call the app navigator to open the task details. (To read more about how navigation works in SwiftUI, go to *Chapter 4*.)

Now, let's see what happens when we run this intent (*Figure 13.7*):

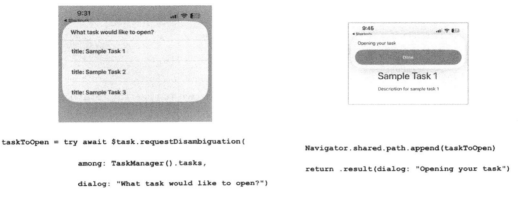

```
taskToOpen = try await $task.requestDisambiguation(
                among: TaskManager().tasks,
                dialog: "What task would like to open?")
}
```

```
Navigator.shared.path.append(taskToOpen)
return .result(dialog: "Opening your task")
```

Figure 13.7: The Open a task intent

Figure 13.7 shows how the Open a task intent looks in the two stages where the task parameter is nil.

First, it opens the app (that's because we set the openAppWhenRun variable to true).

Then, it displays a native dialog where the user can pick a task. Notice that the task display name (title:<title of task>) is something we defined in the displayRepresentation variable when we conformed to AppEntity (in the *Conforming to AppEntity* section).

Later, we navigate to our task details screen and notify the user by returning a dialog with a corresponding message.

Letting the user pick a task to display is a nice use case, but that's not where the real power of intents is.

Let's try and integrate the Open a task intent into another intent by chaining them together.

Chaining app intents

Let's go back to AddTaskIntent, which we created in the *Adding a parameter to our app intent* section, and examine its perform() function:

```
func perform() async throws -> some ReturnsValue<String> {
        TaskManager().addTask(Task(title: title))
        TaskManager().saveTasks()
        return .result(value: title)
    }
```

The return type in the `perform()` function is `ReturnsValue<String>`. Let's modify this function to return a `Task` instance:

```
func perform() async throws -> some ReturnsValue<Task> {
      let newTask = Task(title: title)
      TaskManager().addTask(newTask)
      TaskManager().saveTasks()
      return .result(value: newTask)
}
```

In the new `perform()` function, we changed only two parts – the return type (now it's `ReturnsValue<Task>`) and the return statement, which now returns our newly created task.

Let's go back to the Shortcuts app and chain `AddTaskIntent` and `OpenTaskIntent` together (*Figure 13.8*):

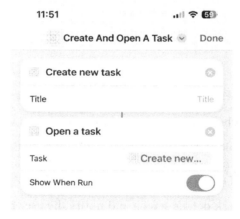

Figure 13.8: A shortcut with Create and Open a task intents

Now we have a shortcut that creates a new task and opens it in the app, and we've done this with very little code!

But what about the properties we defined as part of the `Task` entity? We haven't used them yet! Let's see how to use them with other intents.

Integrating our intent to other intents

We've seen how to chain the added task with the `Open a task` intents, but that was straightforward – we created both intents, so both were aware of the `Task` entity. But what do we do when we need to return the `Task` entity to an intent from another app developer? The first option is to select one of the properties.

Selecting one of the properties

One of the good things about `AppEntity` is that it creates a structure that can be utilized across our system:

```
struct Task: Identifiable, Codable, AppEntity {
        static var typeDisplayRepresentation: 
TypeDisplayRepresentation { .init(stringLiteral: "Task") }


    @Property(title: "Title")
    var title: String

    @Property(title:"Description")
    var description: String
```

Our `Task` structure contains a display name (`Task`) and two properties – `Title` and `Description`. We can use that to pass on one of these values to the next action in the Shortcuts app.

For example, let's say we want to create a new task and send its title in a message. Because we defined the `title` variable as an `AppEntity` property, it will show up in the Shortcuts app (*Figure 13.9*):

Figure 13.9: Choosing the Title property in the Shortcuts app

Figure 13.9 shows how we can select one of the `AppEntity` properties to pass on to the *send message* intent.

Passing on `Title` is obvious—`Title` is a string, and we can easily use it as an input for other actions. But what if we used `Task` as an input for the *send message* action? We did that in the *Chaining app intents* section, but that was between two actions from the same app. Can we share an entity between two different apps?

That's why we have the **Transferable** protocol. Let's use it!

Use the Transferable protocol to pass the entire entity

Let's step outside the framework of `AppIntent` for a second. The idea of sharing data is not limited to `AppIntent`—we have more use cases when we need to share data. For example, dragging and dropping between views or even between apps is one example of sharing data. Another example would be copying and pasting between screens or apps.

The main challenge when performing sharing is finding a data type each app agrees on.

To address that sharing problem, Apple introduced the `Transferable` protocol in iOS 16, making sharing data between apps or different spots easy.

Transferable's main usages are copying and pasting and dragging and dropping, but it is also great for sharing app entities in the Shortcuts app.

Now, let's extend `Task` to conform to `Transferable`:

```
extension Task: Transferable {

    static var transferRepresentation: some TransferRepresentation {
        ...
    }
}
```

The `Transferable` protocol has one static variable called `transferRepresentation`. This variable allows us to define how the structure is represented when sharing it with different apps or views.

When working with the `AppIntent` framework, we have several ways to fulfill the `transferRepresentation` variable:

- `DataRepresentation`: We use `DataRepresentation` to convert our object to a data format such as RTF or an image PNG
- `FileRepresentation`: We use `FileRepresentation` to export our entity as a file, such as a PDF
- `ProxyRepresentation`: This provides an alternative in case none of the other representations are suitable

Let's see how we can support both RTF and plain text:

```
extension Task: Transferable {

    static var transferRepresentation: some TransferRepresentation {
        DataRepresentation(exportedContentType: .rtf)
        { task in
            task.asRTF()!
        }

        ProxyRepresentation(exporting: \.title)
    }
}
```

In our code example, we used both `DataRepresentation` and `ProxyRepresentation` to support both RTF and plain text.

This means that when we try to share the `Task` entity, the `Transferable` mechanism will try to export the `RTF` first and turn to `ProxyRepresentation` as a fallback.

In addition, in the Shortcuts app, the user can select the data type they want to be exported to the next step in the script (*Figure 13.10*):

Figure 13.10: The Shortcuts app with different data formats

Figure 13.10 shows how the user selects the data format for the exported item.

The more formats and data types we support in our app entities, the more choices our users have for using our data.

At this stage, we know how to export our app intents and entities. In a way, the system knows what our app is capable of. Let's see how we can take this one step further and adjust it to work with Apple Intelligence.

Adjusting our app intents to work with Apple Intelligence

In the previous chapter, we discussed how we can take advantage of some of iOS's machine learning and AI capabilities. One of the things that evolved in that area in iOS 18 is Siri. Siri is now smarter than ever and can allow users to perform tasks in natural language.

For example, the user can say something like *Send this photo to my mom* to Siri, and Siri can handle that task without needing the exact phrase.

Siri's new capabilities are exactly where our app intents come in. Imagine that on one side, we have Siri, which can understand the user intent. On the other hand, we have the various actions we expose to the system. So, we must find a way to bind between the user intent, as Siri understands it, and our app actions. This is what we call an **Assistant Schema**.

Exploring the Assistant Schema

The Assistant Schema idea is simple yet advanced and full of potential.

Let's look at *Figure 13.11*, which describes how Assistant Schema works:

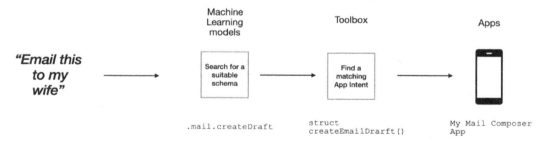

Figure 13.11: The Assistant Schema flow

Figure 13.11 describes how the Assistant Schema flow works:

1. The user requests Siri in a *free and natural phrase* for a specific action. In this case, the user says, "Email this to my wife."

2. Siri and Apple Intelligence convert the request to one of the system's predefined schemas using sophisticated machine learning models. In this case, Apple Intelligence translates the user request to the `createDraft` schema from the `mail` domain.
3. Now that Siri knows the selected schema, it looks to find a matching intent in its toolbox. We can associate some of our app intents with a specific schema.
4. Siri launches and performs a corresponding app intent according to the decision taken in the previous step (the toolbox step).

The fact that Apple Intelligence takes responsibility for understanding our users and picking the right action makes our job as developers simple – all we need to do is ensure our intents match a predefined Assistant Schema.

Let's see how to do that. Imagine we have an amazing mail client app that we've built. Users can browse their email accounts using our app and create and send new email messages.

One of the app's main actions is creating a new draft; so, we've created an app intent to expose that capability:

```swift
struct SendDraftIntent: AppIntent {

    static var title: LocalizedStringResource { "Send new email" }

    @Parameter(title: "Body")
    var body: String?

    @MainActor
    func perform() async throws -> some ReturnsValue
    <MailDraftEntity>{
        let mailDraftEntity = MailDraftEntity(body: EntityProperty(title: LocalizedStringResource(stringLiteral: body!)))
        ComposeDraftManager.shared.isPresentingCompose = true
        return .result(value: mailDraftEntity)
    }
}
```

In this code example, we created a `SendDraftIntent` app intent that takes a body variable, creates a mail draft entity, and launches the mail composer.

It's great that we have a *create draft* action our users can add to their shortcuts. However, Siri doesn't care that we called our app intent `SendDraftIntent` – we want our intent to be part of its Apple Intelligence toolbox. In other words – add it to the Assistant Schema.

To do that, we add a special Swift macro called `AssistantIntent(schema:)`:

```
@AssistantIntent(schema: .mail.createDraft)
struct SendDraftIntent: AppIntent {
```

Adding the `AssistantIntent` Swift macro with the `.mail.createDraft` schema to our app intent signals Apple Intelligence. This is a *create draft* app intent regardless of how we named it.

> **Swift Macros – a reminder**
>
> Swift Macros is a new feature Apple added to iOS 17 that can inject new properties and functions and manipulate our code to various needs. To learn more about Swift Macros, refer to *Chapter 10*.

The `.mail.createDraft` schema is built of two parts – the domain (`mail`) and the schema (`createDraft`). We can use more schemas in the `mail` domain, such as `deleteDraft`, `saveDraft`, or `replyMail`. In addition, we have more domains and schemas to work with, such as presentations, payments, browsing, photos, books, and more. To see the full list of domains and schemas, go to Apple's documentation at https://developer.apple.com/documentation/AppIntents/app-intent-domains.

So, what does the `AssistantIntet` Swift macro do?

The first thing it does is add a new static variable named `__assistantSchemaIntent`:

```
static let __assistantSchemaIntent = AssistantSchema(.mail.createDraft)
```

This static variable marks our intent as a `createDraft` schema. It also ensures that our intent conforms to the `AssistantSchemaIntent` protocol, which gives it more capabilities.

Once that happens, it's time to adjust our code according to what the compiler requires:

- We can remove the `title` static variable, as the App Intents frameworks implement it for us.
- The same goes for the `@Parameter` argument. The App Intents framework implements that for us, so we can also remove that.
- We must add more properties to our app intent that are part of the `createDraft` assistant schema – `account`, `attachments`, `to`, `cc`, `bcc`, and `subject`.

Our new modified `SendDraftIntent` now looks like this:

```
@AssistantIntent(schema: .mail.createDraft)
struct SendDraftIntent: AppIntent {
    var account: MailAccountEntity?
    var attachments: [IntentFile]
    var to: [IntentPerson]
```

```
    var cc: [IntentPerson]
    var bcc: [IntentPerson]
    var subject: String?
    var body: String?

    @MainActor
    func perform() async throws -> some ReturnsValue
    <MailDraftEntity>{
        let mailDraftEntity = MailDraftEntity(body:
          EntityProperty(title: LocalizedStringResource
            (stringLiteral: body!)))
        ComposeDraftManager.shared.isPresentingCompose =
          true
        return .result(value: mailDraftEntity)
    }
```

In this code example, we can see the new modified version of the `SendDraftIntent` struct. The new properties, such as `attachments`, `to`, `cc`, and `bcc`, have particular types, such as `IntentFile` and `IntentPerson`. The `AppIntent` framework uses this type to identify people and files and have a clear interface that the system can work with. Besides adding them to the `SendDraftIntent` struct, we don't need to do anything with them except use them in our `perform()` function.

When we look at the code, one question arises: How do we know what properties to add for each domain and schema?

At this time of writing, there is clear documentation of what properties each schema requires. However, adding the `AssistantIntent` Swift macro and building the project creates new errors that provide information about the missing information.

One exception, though, is the `account` property, which requires us to declare an `AssistantEntity`-based struct. Let's discuss it.

Creating AssistantEntity

When we discussed `SendDraftIntent`, we reviewed several properties, such as `attachments`, `to`, and `bcc`. We saw that for each one, the `AppIntents` framework provides a dedicated type, such as `IntentPerson` and `IntentFile`.

The case of the `account` property is a little bit different:

```
    var account: MailAccountEntity?
```

`MailAccountEntity` is not part of the framework of `AppIntent`—it's a type we define that must fulfill the Assistant Schema requirements, similar to what we did in `SendDraftIntent`. Let's see how to implement it:

```
@AssistantEntity(schema: .mail.account)
struct MailAccountEntity {
    let id = UUID()
    var emailAddress: String
    var name: String

    static var defaultQuery = AccountQuery()

    struct AccountQuery:EntityStringQuery {
        func entities(matching string: String)
          async throws -> [MailAccountEntity] {
            []
        }

        init() {}
        func entities(for identifiers: [MailAccountEntity.ID])
          async throws -> [MailAccountEntity] {
            []
        }

    }

    var displayRepresentation: DisplayRepresentation
      { DisplayRepresentation(stringLiteral: name) }
}
```

In this example, we can see that our `MailAccountEntity` struct has a Swift macro named `@AssistantEntity(schema: .mail.account)`. This macro makes our entity conform to `AssistantSchemaEntity` and requires the struct to implement important properties, such as `emailAddress` and `name`.

The Swift macro also requires us to add a default query to help the system fetch and locate accounts when needed.

The second entity we need to implement is `MailDraftEntity`:

```
@AssistantEntity(schema: .mail.draft)
struct MailDraftEntity {
    static var defaultQuery = Query()
```

```
        struct Query: EntityStringQuery {
            init() {}
            func entities(for identifiers: [MailDraftEntity.ID])
              async throws -> [MailDraftEntity] { [] }
            func entities(matching string: String)
              async throws -> [MailDraftEntity] { [] }
        }

        var displayRepresentation: DisplayRepresentation
          { DisplayRepresentation(stringLiteral: "\(subject ?? "")") }

        let id = UUID()

        var to: [IntentPerson]
        var cc: [IntentPerson]
        var bcc: [IntentPerson]
        var subject: String?
        var body: String?
        var attachments: [IntentFile]
        var account: MailAccountEntity
    }
```

`MailDraftEntity` contains properties such as those of `SendDraftIntent`. That's because it's the result of the `SendDraftIntent perform()` function, and Siri can use it to chain the information to other actions in its toolbox.

Adding both `MailDraftEntity` and `MailAccountEntity` can be annoying – it requires us to adjust our information to a specific interface. However, doing that makes our Siri integration flawless and effective.

Once we have everything set, the user can see a photo and say something like, "Email this photo using MyMailComposer app," and Siri will launch our app with a new draft.

> **An important disclaimer about the code snippets of this section**
>
> Apple Intelligence has not yet been rolled out as of the time of writing this book. This means the code was successfully compiled but has not yet been tested to work with Apple Intelligence. When Apple Intelligence reaches your region, you may need to make some adjustments to your code in order for it to work with Siri as expected.

As one of Apple's senior managers once said, we should consider all of our app actions to be app intents. This approach makes the possibilities for users to interact with our app limitless.

Summary

That was an exciting chapter! This is not only because app intents are a very exciting topic but also because it's the first time we're truly integrating our code with one of Apple's most significant technologies.

In this chapter, we discussed the concept of app intents, created a simple app intent with different use cases, formalized our content using app entities, and even adjusted them to work with Apple Intelligence. By now, we should be ready to bring Siri to our app in no time!

The next chapter looks at our app from a different perspective – quality.

14
Improving the App Quality with Swift Testing

Why is testing part of a coding book? Isn't testing part of the **Quality Assurance** (**QA**) team's remit?

You will soon discover that testing is part of our development cycle and our culture as iOS developers. Many developers see testing as an essential task that they don't have time for. Unfortunately, they pay the price later with bugs and long refactors.

In this chapter, we will do the following:

- Understand the importance of testing
- Learn the testing history of Xcode
- Explore the Swift Testing framework basics
- Understand how to manage tests with suites, test plans, and *Schemes*
- Learn tips that can help us maintain our tests

By the end of this chapter, you will be ready to leverage your testing skills with Swift Testing.

Before we answer the *how* question, let's start with the *why*.

Technical requirements

You must download Xcode version 16.0 or above for this chapter from Apple's App Store.

You'll also need to run the latest version of macOS (Ventura or above). Search for Xcode in the App Store, and select and download the latest version. Launch Xcode, and follow any additional installation instructions that your system may prompt you with. Once Xcode has fully launched, you're ready to go.

This chapter includes many code examples, some of which can be found in the following GitHub repository: https://github.com/PacktPublishing/Mastering-iOS-18-Development/tree/main/Chapter14

Understanding the importance of testing

For many developers, testing is an unnecessary overhead they must deal with when writing code.

This way of thinking is somehow understandable. We've finished writing our code, built an application, and seen that everything runs as expected. Instead of moving to our next task, we need to change the target, adding a test function just so we can see again that it works fine. Why waste our time on it?

Also, in many cases, writing these test functions takes a lot of work. How can we test a SwiftUI view or a network call? What does it even mean?

These all summarize why testing is not a common practice, or at least not enough.

The root of this problem is how developers approach testing and writing code in general. Testing is more than checking whether our functions run as expected; it's about code structure, separation of concerns, the writing process, working culture, and how we treat our day-to-day jobs.

Let's look at the following function:

```
func canUserAddTask(to list: List, user: User) -> Bool {
    if list.isLocked {
        return false
    }
    
    if !list.allowedRoles.contains(user.role) {
        return false
    }
    
    return [.privateList,
       .publicList].contains(list.sharingAttribute)
    
}
```

This function checks whether a user can add a task to a specific list based on criteria, such as permissions, list type, and status. Now, imagine we need to ensure that this function works properly. How can we do that? Do we need to run our app in different states to see the results?

We all know that ensuring our code runs correctly is part of our development process. This is a classic example of how writing test cases and running an app in different states can ease our development process. We understand why testing is so important when adding future tasks such as refactoring and bug fixes.

Before we delve into Swift Testing, let's understand the testing history in Apple platforms.

Learning the testing history in Apple platforms

As Apple development tools evolved over the years, the testing tools have also developed.

The first dedicated testing framework for Apple platforms was **SenTestingKit**, based on the OCUnit open source framework.

SenTestingKit was introduced in 2005 and integrated into Xcode, providing basic functionality for writing and running Objective-C code.

In 2013, Apple introduced **XCTest**, which takes a more modern approach to testing, with better Xcode integration and support for Objective-C and Swift.

Let's take the code example in the *Understanding the importance of testing* section and see an example of an XCTest test:

```
class CanUserAddTaskTests: XCTestCase {

    func testCanAddTaskWhenListIsLocked() {
        let list = List(id: "1", isLocked: true,
           sharingAttribute: .privateList, allowedRoles:
           [.admin, .member])
        let user = User(role: .admin)
        XCTAssertFalse(canUserAddTask(to: list, user:
           user), "User should not be able to add a task
           when the list is locked")
    }
}
```

In this user example, we see a simple test function that tests whether a user can add a task to a locked list.

There are a few things worth noting:

- The test function is part of the `CanUserAddTaskTests` class, inherited from the `XCTestCase` superclass.
- The test function name starts with the `test` phrase. The `test` phrase indicates the XCTest framework, which is a testing function.
- The test validation expression is done by a specific function (`XCTAssertFalse`) that checks whether a particular expression is `false`. We have a list of functions for various conditions.

While these are all part of how we write tests in Xcode, they are not aligned with the modern Swift/SwiftUI approach – working with structs, macros, and more simple and generic Swift functions. That's where Swift Testing comes into the picture.

Let's explore Swift Testing together.

Exploring the Swift Testing basics

We will start our journey by adding the Swift Testing framework to an existing project.

Select **File | New | Target** from the Xcode's menu to do that. Then, in the template chooser, locate **Unit Testing Bundle** and select it (*Figure 14.1*):

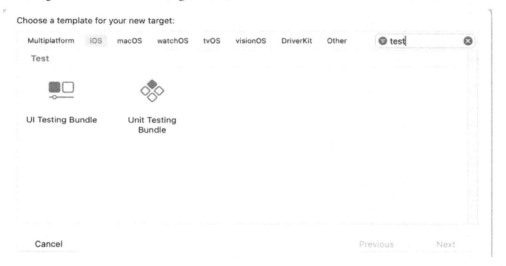

Figure 14.1: The new target template chooser

Figure 14.1 shows the template chooser window in Xcode. When performing a search for testing, **Unit Testing Bundle** is easy to locate. Note that we also have a **UI Testing Bundle** template. However, UI testing isn't supported yet in Swift Testing, so we'll focus now on unit testing.

> **How can we perform UI testing?**
> UI testing, also known as end-to-end testing, is a different topic in app testing. It is also what we call "black-box" testing, meaning that the test function doesn't know the internal code, only the user interface components. The basic way to conduct a UI test is to use XCTest, Apple's previous testing framework. However, there are services that provide simpler or multi-platform ways of running UI tests remotely.

Once you select the **Unit Testing Bundle** template, hit **Next**. Now, we'll need to fill in some details about our new test target (*Figure 14.2*):

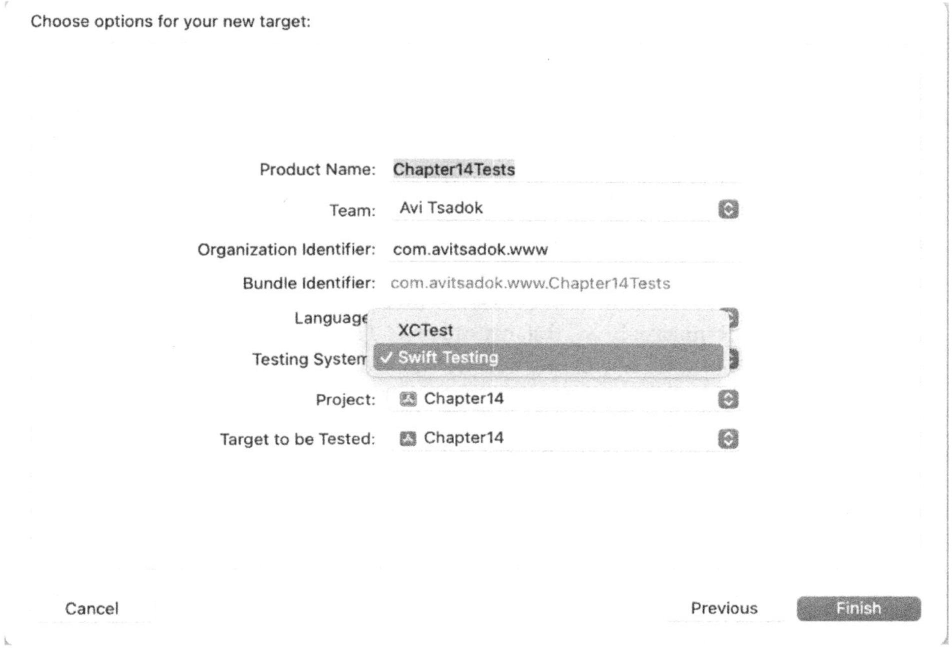

Figure 14.2: Choosing options for our new test target

In *Figure 14.2*, we can fill in the target's name, team, and bundle identifier. We can also choose between the old XCTest and the new Swift Testing frameworks. In this case, we'll select **Swift Testing**.

Hit **Save**, and congratulations – you have a new testing target!

Let's write our first test!

Adding a basic test

Our template comes with a basic, empty test function:

```
import Testing
struct Chapter14Tests {
    @Test func testExample() async throws {
        // Write your test here and use APIs like
        `#expect(...)` to check expected conditions.
    }
}
```

Even though the code is very minimal, we can see a couple of changes compared to XCTest:

- **Import testing**: The Swift Testing framework's namespace is `Testing`, and we should import it into every file we want to test.
- **Working with structs**: Unlike XCTest, which requires creating a class that conforms to `XCTestCase`, we work with structs in Swift Testing. **Structs** are not only lighter and easier to use but also more helpful when we try to run our tests in parallel. Remember that structs are value types, meaning that each time we pass a struct, we get a copy of the data. This helps when trying to check states when testing.
- **Use of the `@Test` Swift macro**: We mark the function with the `@Test` macro, which helps the SwiftData framework manage its tests.
- **Use of the `#expect` macro**: In XCTest, instead of `XCTAssert` functions, we use the `#expect` macro, which is helpful for any expression we want to test.

We can run our test quickly by tapping the diamond button next to the test function or pressing ⌘U. If everything works as expected, our test should pass.

Now, let's fill our code with some actual tests. In our example, we have a view model that handles a counter. We have `increment` and `decrement` functions and a `count` variable:

```
class CounterViewModel: ObservableObject {
    @Published var count: Int = 0
    func increment(by value: Int) { }
    func decrement(by value: Int) {}
    func reset() {}
}
```

Let's test the `CounterViewModel` functionality using Swift Testing.

The first thing we need to do is provide Swift Testing with access to our app target:

```
@testable import Chapter14
```

We add the `@testable` attribute to the `import` command to enable access to internal entities.

Now, let's write our first test function:

```
    @Test func testViewModelIncrement() async throws {
//        preparation
        let viewmodel = CounterViewModel()
        viewmodel.count = 5
        
//        execution
        viewmodel.increment(by: 1)
```

```
//          verification
            #expect(viewmodel.count == 6)
    }
```

In our test function, we initialize the view model, call its increment function, and verify the results. The test fails if the expression inside the `#expect` macro function is `false`.

These three stages – preparation, execution, and verification – are part of any test flow, regardless of whether we use Swift Testing or any other testing framework.

Now, let's rename our struct (which contains this test) `CounterViewModelTests` and run our test.

In Xcode, we can open the left pane on its tab (or just press ⌘6), and then we can see our test list (*Figure 14.3*):

Figure 14.3: The tests listed in Xcode

In *Figure 14.3*, we can see the structure of our tests on the testing pane, which is reflected in the way we create our struct and test functions.

At the beginning of this chapter, we discussed the differences between Swift Testing and Xcode by examining a simple code example. One of these changes was the usage of the `@Test` macro.

Besides indicating a test function, the `@Test` macro has additional features to help us configure our tests.

For example, let's use the `@Test` macro to provide a name to our test function.

Providing names to our test functions

Providing expressive and meaningful names to test functions is crucial and can be valuable when we have hundreds of tests in our project.

To do this in XCTest, we need to rename the test function to something like this:

```
func testViewModelIncremenetFunction_incrementBy1_accept5_expect6
```

The function name describes the test correctly, but it feels cumbersome and awkward, especially when we have hundreds of test functions.

With the `@Test` Swift macro, we can provide a readable name for each test:

```
@Test("Test the increment function. Accepts 5 and expect 6. ") func
testViewModelIncrement()
```

Adding the test description to the `@Test` Swift macro makes it much more readable, and it also integrates nicely with Xcode (*Figure 14.4*):

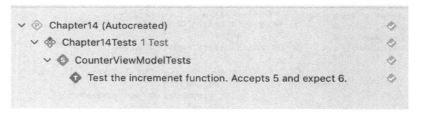

Figure 14.4: The test pane in Xcode, with a custom name

Figure 14.4 shows the same test function as before, now with a readable and meaningful name.

The `@Test` Swift macro provides much more than just naming our functions. We can also use it to disable and enable tests. Let's see how to do that.

Enabling and disabling tests

Sometimes, a test can become irrelevant, and we want to remove it from our test list temporarily. We can delete it or comment on it. However, these solutions may need to be more comfortable and practical in the long run. So, let's use the `@Test` macro to make that more elegant.

In Swift Testing, all tests are enabled by default. To disable a specific test, we can use the `disabled()` function:

```
@Test("Test the incremenet function. Accepts 5 and expect 6. ",
.disabled()) func testViewModelIncrement()
```

We can see that the `disabled()` function is now one of the `@Test` parameters. In this case, the test function won't run, and we can also see that the function is now disabled in the test pane (*Figure 14.5*):

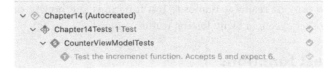

Figure 14.5: A disabled test in the test pane

Figure 14.5 shows our test function grayed out. In this case, performing an entire test run will skip that test.

However, there are cases where we need our test function to run only under specific conditions, such as when a user is logged in or in one particular A/B test condition.

In this case, we will implement the condition within the test function as a guard statement, which makes the test function succeed. But that doesn't sound like a good solution – having a test function succeed when it's not running.

Fortunately, enabling a test function based on specific criteria is a feature Swift Testing supports. All we need to do is add the enabled function within the `@Test` macro head, including a Boolean expression:

```
@Test("Test the decrement function.", .enabled(if: AppSettings.
CanDecrement)) func testTheDecrementFunction() async throws
```

In this code example, we see a new test function called `testTheDecrementFunction`. We added a condition to the test function that would run only if we enabled the ability to decrement in the app settings. In this case, the `AppSettings.CanDecrement` expression returns `false`. Therefore, Swift Testing skips the test function at runtime.

When using the enabled function, precisely defining the test goal is essential. For example, when using `AppSettings`, we may want to test the results of the decrement function when the feature is turned off. We need to disable tests according to a Boolean expression only when it's clear that the function is irrelevant under specific conditions.

If we try to run a test when the `enabled()` function returns `false`, we'll see something like *Figure 14.6*:

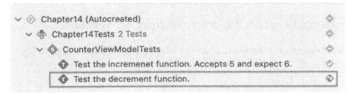

Figure 14.6: A skipped test function due to a specific false condition

In *Figure 14.6*, we can see that the test function is not grayed out, as in the case of using the `disabled()` function. However, it wasn't running, and we can also see the skipped icon on the right.

We have seen how to provide readable names to test functions and how to disable or enable tests. Now, let's discuss another excellent Swift Testing feature – *tags*.

Tagging our test functions

Generally, we group tests according to our project structure. For example, we could create a structure of test functions for a specific class or a structure. Another example would be to create a test structure for a particular feature or service. However, there are additional ways we can organize our test functions. We could arrange them according to priority – critical or sanity tests – or according to their system levels, such as UI or business logic layers.

Instead of finding workarounds for that organization problem, Swift Testing provides an organization feature called **tags**.

We'll start by defining a new tag in the test bundle:

```
extension Tag {
    @Tag static let critical: Self
}
```

We extended the `Tag` structure in this code and added a new static variable, named `critical`.

We can define and use as many tags as we want across our bundle. Therefore, it is a best practice to manage all our tags in one place and a separate file.

Now that we have a new tag, let's add it to one of our tests:

```
@Test("Test the reset function", .tags(.critical))
    func testResetFunction() {
```

In this code example, another `@Test` macro function, called `tags()`, provides the new `critical` static variable we created in the previous code example.

Note that we can provide multiple tags to the same test function:

```
.tags(.critical, .calculations, .performance))
```

In this example, we marked a specific function with three different tags.

The ability to mark a function with multiple tags can be powerful, as it provides flexibility with our tests' organization.

One thing is missing here – even though tagging functions look lovely, we haven't discussed how to actually use our tagging.

Let's look at *Figure 14.7*:

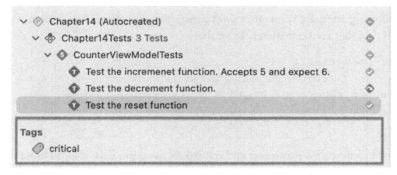

Figure 14.7: The Tags section of the test pane in Xcode

Figure 14.7 shows a new section called **Tags** in the Xcode test pane. The **Tags** section now includes the `critical` *tag* we defined for our `reset()` function in the last code example. Xcode scans our tags' usage and organizes them for us. This is how deep the Swift Testing integration with Xcode is.

Now that we have all our tags listed, we can run all our critical tests (*Figure 14.8*):

Figure 14.8: Running all critical tests

In *Figure 14.8*, the run button is on the right. Tapping it will run all our critical marked tests.

Now, for the practical usage of tags in testing, working with tags is similar to how tagging works in other places. When we group tests in files, we usually do that by *concern* – a layer, service, module, and so on. Conversely, tagging helps us group tests by their *types* (sanity, smoke or regression, integration, or unit) or a *property* (a priority, for example).

> **What are smoke tests?**
>
> We write **smoke tests** to check a system's operations by testing basic functionality. While they may sound like a sanity test, they are much lighter and faster than that. For example, we can try to perform a login and a basic data sync, and the results can indicate whether we have a severe problem with our app or the backend.

Working methodologically with the tagging system can enhance our testing and open new possibilities.

Our `@Test` macro features list doesn't end with tagging. Let's examine a Swift Testing feature that can save us a lot of time – *arguments*.

Working with arguments

Imagine the following scenario. We wrote a Swift function that performs a very clever calculation – for example, a function that converts meters to yards:

```
struct UnitConverter {
    static func metersToYards(_ meters: Double) -> Double {
        return meters * 1.09361
    }
}
```

Our function takes a `meters` parameter and returns its value in yards. It looks like a straightforward function, but we must perform several tests to see whether it works as expected.

So, let's write tests for this function:

```
struct UnitConverterTests {

    @Test func testConvertingMetersToYards_1meter() {
        #expect(UnitConverter.metersToYards(1.0) ==
            1.09361)
    }

    @Test func testConvertingMetersToYards_3_5meter() {
        #expect(UnitConverter.metersToYards(3.5) ==
            3.827635)
    }
}
```

In this code example, we wrote two test functions that perform the same test but with different parameters. They even have very similar names. Even though this solution works fine, it doesn't scale up very nicely. What if we want to test 10 different variants or parameters? And what if we need to change the function name we are testing?

One option is to perform one test function that contains all of the different options:

```
@Test func testConvertingMetersToYards () {
    #expect(UnitConverter.metersToYards(1.0) == 1.09361)
    #expect(UnitConverter.metersToYards(3.5) == 3.827635)
}
```

We created one test function with two `#expect` statements in this code example. That will probably work; however, managing and monitoring them is more challenging now that we have both statements in one function.

To solve that, Swift Testing has a feature named **arguments**, which allows us to run our tests with different values repeatedly.

Let's see that in action:

```
@Test(arguments: [(1.0, 1.09361), (3.5, 3.827635)])
    func testConvertingMetersToYards(data: (Double,
      Double)) {
        #expect(UnitConverter.metersToYards(data.0) ==
          data.1)
    }
```

This code example may look a little cumbersome, but it is straightforward. We performed three changes here:

- We added the arguments *parameter* to the @Test macro, which contains an array of tuples. Each tuple represents a few meters and its corresponding number of yards. For example, the (1.0, 1.09361) tuple represents a conversion between 1 meter and 1.09361 yards. This array is the list of test variants we are going to do.

- We added a *new tuple parameter called* data *to our test function.* With each test run, Swift Testing passes a tuple from the arguments list to the function using this parameter. The parameter type must be aligned with the argument type.

- In the #expect macro, we now *compare the two tuple values* instead of fixed sizes, like in the previous examples.

The term *arguments* can be misleading. In the context of testing, it means that arguments allow us to run our code in different use cases and states.

And if passing all the different use cases within the @Test macro is cumbersome, we can store them in a separate variable:

```
let convertingTests: [(Double, Double)] = [(1.0, 1.09361),
                                            (3.5, 3.827635)]

struct UnitConverterTests {

    @Test(arguments: convertingTests)
    func testConvertingMetersToYards(data: (Double,
      Double)) {
        #expect(UnitConverter.metersToYards(data.0) ==
          data.1)
    }
}
```

In this code example, we moved our use cases into a dedicated constant for better readability.

If we look at the Xcode testing pane again, we can now see a list of our use cases and their states (*Figure 14.9*):

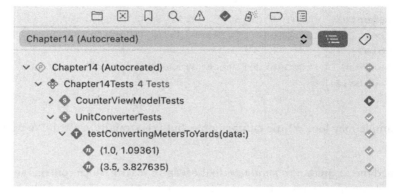

Figure 14.9: Argument testing in the Xcode testing pane

Figure 14.9 shows why argument testing in Swift Testing is so powerful. Instead of having several test functions in the list, we can see one with several use cases.

Argument testing adds another layer to our testing, something we don't have in XCTest.

> **Why doesn't XCTest support parametrized testing?**
>
> Using attributes to perform parametrized testing is not new in the testing world. Most testing frameworks support adding arguments to their test functions out of the box. However, even though it is possible to perform parametrized tests in XCTest, it requires creating several test functions that call a central function that performs the actual test. This is an ad hoc and unnatural solution. The reason is that Apple wanted to create a simple testing framework, and locating the test function in XCTest works according to a simple function signature (functions that start with the phrase `test`). Adding arguments made the locating process complex.

Now that we have reviewed the Swift Testing basics, let's see how to manage our tests.

Managing our tests

Anyone who has previously worked with tests knows that writing tests is one thing and managing them in the long run is another.

If you don't have testing experience, you might think that simply running all your tests one after the other is sufficient. But down the road, things become much more complex – different configurations, environments, and even test goals – all translating to a need for a more robust testing management system.

Before managing our testing system, let's review our Xcode testing structure.

Going over the testing structure

So far, we have discussed how to write testing functions, but besides grouping them in structures, we haven't discussed anything related to managing them.

A whole set of tools can help us manage our test efficiency in Xcode. Let's review the different blocks that can help us adapt a flexible system to our needs:

- **A testing suite**: A testing suite can group several testing functions and child suites.
- **A test plan**: A test plan groups different test functions and test suites. It can include or exclude test functions marked with tags. But it doesn't stop there – test plans can run multiple times in different configurations with different data and environments. This is a powerful tool that can help scale up our testing strategy.
- **A Scheme**: Inside each *Scheme*, we have several build options. One is **Test**, where we must describe what will happen when testing that specific *Scheme*. In the **Test Build** option, we can define precisely what test plans we will run and on which target.

When we look at the different testing building blocks, we can see that the testing structure is complex and requires some thinking.

Let's try to explain the hierarchy by examining *Figure 14.10*:

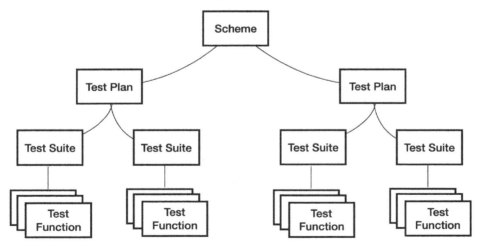

Figure 14.10: Relations between the different building blocks of testing

Figure 14.10 shows the relations between the different building blocks of testing. Next, we will learn how to build them together, starting with test suites.

Grouping our test functions into test suites

The first building block we are going to discuss is the **test suite**. In fact, we have already built a test suite in this chapter:

```
struct UnitConverterTests {
    @Test func testConvertingMetersToYards_1meter() {
        #expect(UnitConverter.metersToYards(1.0) ==
            1.09361)
    }
}
```

Do you remember this code example? We wrote it in the *Working with arguments* section and created a similar test suite for earlier examples. So, yes, the struct containing our test functions is considered to be a test suite, and Swift Testing recognizes and displays this in the test pane.

However, we can annotate a test suite with the `@Suite` attribute for better customization. Let's add it to our latest test suite:

```
@Suite("Unit converter tests")
struct UnitConverterTests {

    @Test func testConvertingMetersToYards_1meter() {
        #expect(UnitConverter.metersToYards(1.0) ==
            1.09361}
}
```

In this code example, we added the `@Suite` swift macro to our `UnitConverterTests` structure and, by doing so, gave it a more readable name.

Let's see what our test suite looks like in the test pane in Xcode (*Figure 14.11*):

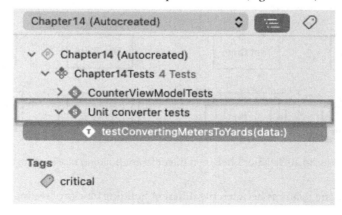

Figure 14.11: The suite in the Xcode test pane

In *Figure 14.11*, we can see our new test suite displayed in the test pane.

If using the `@Suite` macro sounds like how we used the `@Test` macro, you are not mistaken; it's the same idea – providing more information by using a macro.

And, just like the `@Test` macro, we can also mark test suites with tags:

```
@Suite("Unit converter tests", .tags(.critical))
struct UnitConverterTests {
}
```

In this code example, we marked our new test suite with the critical tag we declared in the *Tagging our test functions* section.

In addition, we can also disable the whole test suite using the same `disabled()` function we used in the *Enabling and disabling tests* section:

```
@Suite("Unit converter tests", .disabled())
struct UnitConverterTests {
}
```

In this code example, we disabled the `Unit converter tests` test suite, and Swift Test will not execute any of its tests in the next test run.

Another neat usage for a test suite is its ability to contain nested test suites:

```
@Suite("Unit converter tests")
struct UnitConverterTests {

    @Suite("From meters to yards")
    struct FromMetersToYardsTests {
    // our test functions
    }
}
```

In this code example, we have a test suite named `From meters to yards`, which is part of a bigger test suite named `Unit converter tests`.

Let's see how this is reflected in the Xcode pane (*Figure 14.12*):

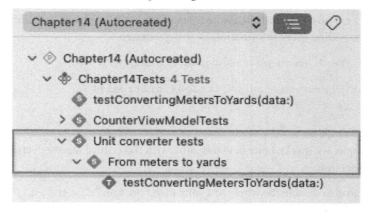

Figure 14.12: Nested test suites in the test pane

Figure 14.12 shows how our new nested test suite is reflected in the test pane. We can also customize the nested suites, such as adding tags and disabling them.

Now that we know how to define a test suite and tags, it is recommended that we remember each feature's role. We use test suites to group different test methods by concern – typically, by writing test functions for a specific class or a structure.

Conversely, we use tags to mark our tests by their type – `critical`, `performance`, `integration`, and so on. If these are the different roles for tags and test suites, what do we do when we want to manage something such as a sanity or a regression test?

That's what *test plans* are for.

Building test plans

To better understand the different testing components, we can think of an app with two layers – the business logic and the UI. The business logic layer is important, but it doesn't describe how a user will use our app – the different use cases and flows.

We must build the UI layer to complete our app, which handles user stories and flows. The business logic is analogous to the different testing suites and functions. These are the building blocks of our testing. However, testing is always in the context of a specific development process.

Let's try to come up with different development processes:

- **Feature development**: We build new features, often adding new classes, structures, and entities
- **Fixing bugs**: We modify existing code
- **Refactoring code**: We modify existing code for better scalability, maintenance, or performance
- **Deployment**: We prepare an app for deployment for QA or production

This is only a partial list of different development processes, but it demonstrates that we are always in the context of a process when we develop.

When we build our testing system, we can describe this process using a test plan. Let's add a new test plan to see how it works.

Adding a new test plan

Test plans are a new feature in Xcode, added to Xcode 11 in 2019. They allow us to pick tests or test suites and run them in a specific configuration and environment. Test plans are our way of expressing how our test functions will be executed.

We always run our tests as part of a test plan. By default, Xcode creates a test plan for us automatically (*Figure 14.13*):

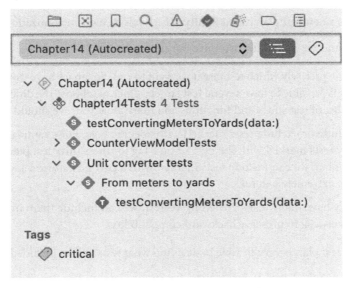

Figure 14.13: The autocreated test plan

In *Figure 14.13*, we can see that Xcode created a test plan for us called *Chapter 14*. To create a new test plan, we can tap the test plan pop-up menu and select **New Test Plan**. After we provide a name for our new test plan, we can see it in our Xcode main pane (*Figure 14.14*):

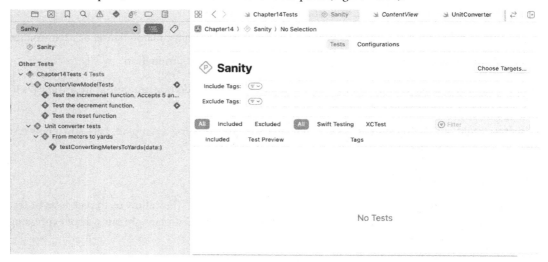

Figure 14.14: The new test plan in Xcode

Figure 14.14 shows a new test plan called **Sanity**, which has its own customization screen.

There are many things we can do to customize our new test plan:

- We can define precisely which test target we want to run. So far, we have worked on a single test target, but it is possible to have several test targets. Once we choose the different test targets, we will see the list of test suites and functions and mark what tests we should include or exclude.
- We can include or exclude tests marked with specific tags. For example, we can choose to include only tests marked with the `critical` tag for the **Sanity** test plan. Alternatively, for a stress test plan, we can include tests marked with a `performance` tag. This is where the tags become extremely helpful.
- If we already have many tests written in XCTest, we can include them in our test plan. This capability is crucial to preserve backward compatibility.

As we can see, the test plan is very flexible in deciding what tests will be included.

However, control over the list of test suites and functions is only a fraction of what we can do with test plans. We can do even more with configurations.

Configuring our test plan

When we started explaining test plans, we said that part of the idea of creating one is defining the environment in which the test plan runs. One example of such an environment is localization – language, region, and location can influence our app in certain use cases.

Trying to simulate an environment for our test functions can be challenging; therefore, test plans have a feature called **Configurations** (*Figure 14.15*):

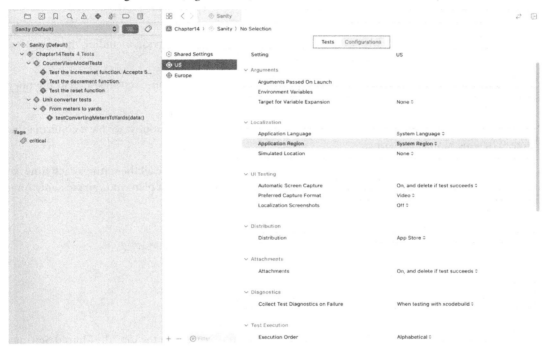

Figure 14.15: The test plan's Configurations tab

Figure 14.15 shows a tab bar at the top of the **Sanity** main pane. The **Tests** tab defines the included tests in the test plan, and the **Configurations** tab defines the different configurations for the test plan.

To add a new configuration, we tap the plus button at the bottom of the window.

A test plan can have many configurations. Each configuration contains a list of settings that can affect our test results. Let's examine them briefly:

- **Arguments**: Each app can run with different Launch and environment variables. We can use them to override our A/B test configuration or define a specific API endpoint. Arguments are powerful tools that help us adjust our app to our needs.

- **Localization**: Language, region, and location are all part of the localization list of settings that we can define. These settings can influence available features, texts, measurement units, and other behavior.
- **UI testing**: If our test plan includes UI tests (not supported yet in Swift Testing), we can decide what happens during screen capturing if there is a test failure.
- **Distribution**: Some APIs can behave differently when running on the App Store than on TestFlight – for example, collecting beta testers' feedback, sandbox issues, and enabling/disabling beta testing features.
- **Test Execution**: Here, we can define the test plan execution behavior, including the execution order, timeouts, and repetition settings.
- **Runtime API Checking, Runtime Sanitization**: Different runtime settings such as memory management, main thread checker, and sanitization.

That's a long list of settings! I felt that when I looked at *Figure 14.15*, but now we have confirmation after reviewing almost each one.

However, the idea behind configurations is that we don't need to redefine all the settings each time we create a new configuration. If you open your Xcode and create a new test plan, you can see something called **Shared Settings** (*Figure 14.16*):

Figure 14.16: Shared Settings

Figure 14.16 focuses on the list of configurations with **Shared Settings** at the top. The **Shared Settings** configuration contains the settings for all configurations unless we explicitly change a specific setting for a particular configuration.

Consider a typical use case – we would probably want the same settings for all configurations except for one or two (e.g., a configuration for different locations or distributions). In this case, we will have the same settings except for the region or the distribution method.

Xcode executes all the configurations in a sequence when running a test plan. However, you can disable a specific configuration by right-clicking on it in the configurations list and selecting **Disable**.

So, let's say we created a sanity test plan and a regression test plan. What do we do from here? How can we tell Xcode what to execute when running tests? This is where the *Scheme* comes into play.

Setting up a Scheme

This chapter is about Swift Testing, not the Xcode build system, but we can't discuss testing and ignore **Schemes**.

Schemes are fundamental to managing our project's build and execution configurations. A *Scheme* defines how our project is built, executed, and tested.

We can write dozens of test functions and create as many test plans as we want, but the bottom line is that when we select **Test** from the Xcode menu or run tests from our CI/CD environment, the *Scheme* defines precisely what will happen.

> **What is CI/CD?**
>
> **CI/CD** stands for **Continuous Integrations/Continuous Deployment**. We use these practices to automate our app build and deploy process. A crucial part of this process is testing – before we deploy a build to TestFlight or the App Store, we want to perform testing to ensure we don't have regressions or other issues. When we build our CI/CD process, we often choose what Scheme to execute.

Looking at the Xcode window, we can locate the *Scheme* name next to the project name. Tapping it will open a list of Schemes where we can change the current *Scheme* or edit it (*Figure 14.17*):

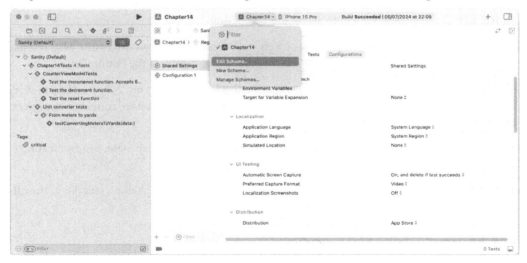

Figure 14.17: Editing the current Scheme

Figure 14.17 shows how to reach the pop-up *Scheme* menu. Tapping on the **Edit Scheme…** option leads us to the **Edit Scheme** screen (*Figure 14.18*):

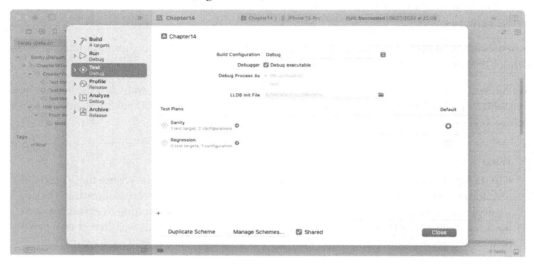

Figure 14.18: The Edit Scheme screen

Figure 14.18 shows that the *Scheme* has six different actions—**Build**, **Run**, **Test**, **Profile**, **Analyze**, and **Archive**. In this screenshot, we will focus on the **Test** action.

Besides choosing the configuration (**Debug** or **Release** in our case), we can determine what test plans to run. We can add an existing or new test plan using the plus button at the bottom left.

That's where we decide what happens when executing the **Test** action on our *Scheme*. Having several *Schemes* configured differently for various purposes can be valuable when we connect our project to a CI/CD system.

For example, we can run a performance test once a month and sanity every night, just by creating two different Schemes that run different test plans.

Now that we know how to create test functions, suites, test plans, and Schemes, let's flip to the other side of the equation and see how to write testable code.

Tips to write testable code

One of the biggest challenges developers face when they try to write tests for code is struggling to write tests for existing functions that could be more testable – for example, functions that contain code that performs network requests or functions that have external dependencies that are difficult to set up.

Writing testable code usually goes hand in hand with writing clean and efficient code. However, we should still follow some writing guidelines if we want our functions to be testable.

Let's explore some of them now.

Writing pure functions

Pure functions are functions that, given the same input, always return the same output and don't rely on external states or have any side effects.

For instance, take the following example:

```
class NumberFilter {
    var numbers: [Int] = []
    var filteredNumbers: [Int] = []

    func filterNumbers(predicate: (Int) -> Bool) {
        self.filteredNumbers =
            self.numbers.filter(predicate)
    }
}
```

This code example contains a `NumberFilter` class with a function called `filterNumbers`. This class performs a predicate on an instance variable and stores the results in another instance variable.

This is a classic example of a non-pure function, since it relies on an external variable and has a side effect. Now, imagine we want to test this function – it requires us to set up a `NumberFilter` instance and set the `numbers` variable. In addition, we need to check the result using the same `NumberFilter` instance, with the `filtersNumbers` instance.

The class can change down the road and may require more setup than before, breaking our test.

Instead, we can make this function pure, like this:

```
func filterNumbers(_ numbers: [Int], predicate: (Int) -> Bool) ->
[Int] {
    return numbers.filter(predicate)
}
```

In the modified example, our function receives the input as a parameter and returns the results as part of its output. This change makes it agnostic to external states and easy to test.

Separating your code based on concerns

As always, a good separation is crucial for our project maintenance (which we will cover in more detail in *Chapter 15*). However, separation is also essential for testing.

The fundamental separation of concerns idea states that each part of our code, whether a variable, function, class, or module, should have one and only one responsibility.

Let's take the following code as an example:

```
func processAndSaveData(_ input: String) -> Bool {
    // Data processing
    let processedData = // <perform some data manipulation
        code>

    // Data saving
    return databaseService.saveData(processedData)
}
```

The `processAndSaveData` function is responsible for two tasks – processing the input data and saving it to the database service.

We can see that the string processing code uses the same function that performs data saving. If we want to test whether the string processing succeeded, we must also ensure that the output has been saved successfully. These two responsibilities are coupled, which makes the code very difficult to test.

To solve that, we can separate the processing code into another function:

```
func processAndSaveData(_ input: String) -> Bool {
    // Data processing
    let processedData = processData(input)

    // Data saving
    return databaseService.saveData(processedData)
}

private func processData(_ input: String) -> String {
    return input.reversed()
}
```

In this example, we gave the processing data task its own function, and now it is possible to test it regardless of the data-saving part.

Our last tip also discusses coupling but, in another context – *protocols*.

Performing mocking using protocols

Sometimes, we don't have a choice but to test functions that reach our network or any external service that can't really simulate during tests.

To overcome that, we can easily create mocks for these services using **protocols**.

Look at the following code:

```
class UserViewModel {
    private let networkService: NetworkServiceProtocol
    var user: User?

    init(networkService: NetworkServiceProtocol) {
        self.networkService = networkService
    }

    func fetchUserDetails(for userId: String, completion:
      @escaping () -> Void) {
        networkService.fetchUserDetails(for: userId) {
          [weak self] user in
            self?.user = user
            completion()
        }
    }
}
```

This code example contains a `UserViewModel` class that fetches user details from the server and stores the results in an instance variable. Testing the `fetchUserDetails` function requires performing a request to the server, which can make our test unstable.

To solve that, we can create a mock class that conforms to `NetworkServiceProtocol` and simulate the network service:

```
class MockNetworkService: NetworkServiceProtocol {
    var userToReturn: User?

    func fetchUserDetails(for userId: String, completion:
    @escaping (User?) -> Void) {
        completion(userToReturn)
    }
}
```

This example demonstrates a mock class that accepts a user's return and can easily mock the whole network process. We achieved that by using a protocol and dependency injection, and we can do the same to store data, authenticate, and so on.

Summary

Testing is crucial to our mission to produce stable, high-quality code. Remember, writing tests is not just a fundamental part of being a professional iOS developer – it is also part of a culture of doing things right.

In this chapter, we've learned about the testing history in Xcode, covered the Swift Testing basics by writing simple tests, learned how to manage our tests using suites, test plans, and Schemes, and even discussed some useful tips to make our code more testable. Now, we should be able to set up a new test plan for our project!

Our following and final chapter, on architecture, touches on some of the principles we discussed here and will also help us create a stable project.

15
Exploring Architectures for iOS

In the previous chapter, we discussed Swift Testing, an essential framework that helps us test our Swift code. App testing is not only a technical topic – it is also a culture. Part of this culture is looking at our project as a set of classes and a whole structure with a certain logic. That's why testing and architecture go hand in hand – they both look at our project as a well-designed system. This holistic approach is essential to meet our product requirements over time.

In this chapter, we will cover the following topics:

- Understanding the importance of architecture
- Learning what exactly architecture is
- Going over the different architectures, such as multi-layer, modular, and hexagonal
- Comparing the different architectures by separations, testing, and maintenance

First, let's understand why architecture is so important.

Technical requirements

You must download Xcode version 16.0 or above from Apple's App Store for this chapter.

You'll also need to run the latest version of macOS (Ventura or above). Search for Xcode in the App Store and select and download the latest version. Launch Xcode and follow any additional installation instructions that your system may prompt you with. Once Xcode has fully launched, you're ready to go.

Understanding the importance of architecture

To understand the importance of architecture, let's try to understand how the iOS development knowledge is built.

Many think that iOS development is centralized around Swift – if we know Swift, we know iOS development.

Nonetheless, iOS development contains many knowledge levels, and the Swift language is only one of them.

Let's try to structure the iOS development to different levels:

- **IDE**: Familiarity with Xcode, its debugging tools, configuration, simulator, builder, and code signing is crucial.
- **Language**: Whether it's Swift, Objective C, or C++, language is a fundamental part of iOS development. It's the basis for daily implementing our app's logic and design pattern.
- **System**: Understanding iOS's unique characteristics, strengths, and limitations is key. Ultimately, we are developing in a particular environment with its own rules and policies.
- **SDK**: The SDK provides the toolset to do whatever we want. SwiftUI, UIKit, Foundation, Core Animation, and many other frameworks are part of the SDK, and with them, we can create beautiful screens with user input components and persistent storage.
- **Design patterns**: These are solutions to common problems and tasks we encounter daily.
- **Architecture**: The high-level organization of our code and project is called its architecture.

We can continue with more knowledge levels – testing, databases, networking, security, and more. The knowledge spectrum has become huge over the years, with more and more capabilities and knowledge required.

Still, many iOS developers don't focus on architecture when they build their apps, and there are some obvious reasons why. For example, developers prefer to see immediate results. Sometimes, it's not only a matter of choice – there are deadlines to meet, and a lack of resources forces us to focus on releasing our features as quickly as possible.

However, ignoring good architectural planning is usually a result of a lack of experience and short-term focus, which gives a clue about how important architecture is.

Let's list some of the influences of good architecture on our project:

- **Maintainability**: Our projects can easily become more extensive and challenging to maintain. A good architecture makes our code base more straightforward to understand and read. It also makes it easier to modify and refactor.
- **Scalability**: The ability to add more features while keeping our project simple and stable is a crucial key to app success. A bad architecture can require a significant overhaul whenever we want to add new features.
- **Flexibility**: A good architecture allows us to quickly change how our app works according to changes in requirements. It also helps us add new features or replace third-party frameworks.

These are some benefits of good architecture, but the picture is clear – we will mainly discuss **long-term influence**. Working hard to create more classes, layers, and protocols in the short term seems like a big hassle. Besides coding, good architecture requires upfront efforts such as good planning, tech design, and a good understanding of paradigms and patterns.

Before we discuss the different types of architecture, let's define what architecture means and what defines good architecture.

Learning what exactly architecture is

Many developers are confused between architecture and what we call a "design pattern". We previously touched on that when we discussed the different layers of knowledge (under the *Understanding the importance of architecture* section), and even though it sounds like a semantic difference, it's crucial to understand the distinction. While architecture refers to the *high-level organization* of our app, such as layers, modules, and components, design patterns are *reusable solutions to common problems*.

To explain that better, let's imagine a building. When planning a private house, we must decide its number of floors, entrance, roof, and garage In short, this is the house's architecture.

In contrast, each floor has its own goal and designation. For example, one floor can be the kitchen and living room, and the second would be the bedrooms. To accomplish that, we need to plan the internal design for each floor, deciding the sizes of the rooms, the door locations, and the different wires and water pipes. In most cases, there are no tricks here – there are standards to follow. These internal designs of the floors can be considered as design patterns – a reusable, specific solution to common problems.

Now, let's go back to our mobile app. We should think of a mobile app's structure as a private house. The data flows in different layers – data, business logic, and **User Interface** (**UI**). We can look at each layer as a different floor in our home. In each layer (or floor), we can use various design patterns to solve other problems. For example, we can use Singleton to manage shared resources or a coordinator to simplify complex navigation needs.

The more design patterns we know, the more solutions we have.

Moreover, let's continue with the house metaphor. In that case, we can come to another conclusion – our choice regarding architecture affects the different design patterns we use for our floors, including the floors' sizes and shapes, or even how they are connected.

So, what are the different types of architecture available, and how do we select an architecture that fits our needs?

Going over the different architectures

Developers make two common mistakes when choosing their project architecture. First, they often say, "*What architecture am I using for my app? MVVM, of course!*"

MVVM is not an architecture – it's a design pattern that aims to solve state and logic management for a particular screen. Not only does it not handle the app structure but it also doesn't even describe how we handle our screens in general. It only describes a particular screen, such as a login or a settings screen.

The second mistake is the idea that we can only choose one of the most common and popular architectures from the list for our project. Most of the architectures you read about are, in fact, a set of principles that can help us decide how to structure our project.

Some principles provide flexibility and decoupling, and some may increase project overhead. We should always consider tradeoffs; these become even more important in architectural designs.

Let's start with the most fundamental architectural idea: the multi-layer architecture.

Separating our project into layers

It's worthwhile to take a moment and discuss two important terms I'm using here. The first is a *project* and not an *app*. The reason for that is that our architectural decisions are related to the whole project – pods, Swift packages, extensions, or even other apps. When we talk about structures, an app is only the expression of our product and how we deploy it.

The second term is using *layers* instead of *tiers* – a typical mistake developers make. When discussing separating a system into tiers, we often refer to hardware separation – different computers, servers, routers, or other hardware components. We should use the term *layers* when discussing separating software such as an app or SDK.

Separating a project into layers, usually three, is a common architectural decision in many projects. The idea is that a basic project has at least three different levels of data and logic handling (*Figure 15.1*):

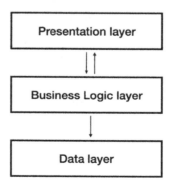

Figure 15.1: The three-layer architecture

Figure 15.1 shows the three layers that we usually separate our app into.

Let's try to understand these layers:

- **Data** can sometimes be called **services**. The Data layer handles the data persistent store, the model entities, network handling, and primarily services that handle data at a low level regardless of the project logic.
- **Business logic** can sometimes be called **domain**. The Business Logic layer handles the app's main logic, including rules and data manipulations.
- **Presentation** handles the UI, user interaction, and navigation.

There are patterns that have even more layers – for example, an *application* layer, which handles the different use cases and can be placed under the presentation layer, or an *infrastructure* layer that handles class extensions, utils, and more.

If you are an experienced developer, the idea of separating a project should be obvious. Separating code creates a testable and maintainable structure that can be scaled over time. However, the idea of working with layers is not always evident.

Ultimately, it comes down to how the data flows around the app.

Controlling the app data flow

Data flow is a central topic in any program. To clarify that term, we must examine how messages and data flow between the different app components. For example, when a user taps the **Save** button on one of their screens, we need to transform that tap into an actual logic decision and continue that to the persistent store, where we can save that information locally. The data flow doesn't end here – at this point, we need to send a message back to the UI that a change has been made in the persistent store, and we should update what's on the screen.

This example demonstrates only a single use case. A standard mobile app may have hundreds of such cases, emphasizing the importance of considering how to divide our project.

Now that we understand data flow, let's discuss the **open** and **closed** layers. In a three-layer architecture, as described in *Figure 15.1*, the presentation layer communicates with the business logic layer. However, does that mean that the presentation layer is also allowed to communicate with the data layer?

For example, the presentation may receive updates about data changes directly from the data layer. Working with business logic as middleware can be more complex and cumbersome in these cases. At that point, we must decide whether our layers are open or closed.

An open layer allows direct interaction between the layer above and underneath. While open layers provide higher flexibility and simplicity, they can also increase coupling and reduce the separation of concerns.

A closed layer enforces strict interaction, and each communication between its adjacent layers must go through the closed layer itself. A closed layer can increase the separation of concerns and loosen the coupling while decreasing flexibility and increasing complexity.

Discussing closed and opened layers might sound a little weird when working with three-layer architecture. The middle layer (business logic) is the only layer that can be either open or closed. However, we can decide whether the layer is strictly or selectively closed.

Each layer is built from a set of components. For example, the presentation layer can be built from the different app screens or flows. The business logic layer can be built from the different logic parts of the app, and the data layer is built from the different services such as network, data, and security.

In some situations, components from the presentation layer must communicate directly with the data layer.

Look at *Figure 15.2*, which shows the basic three-layer architecture of a messaging app:

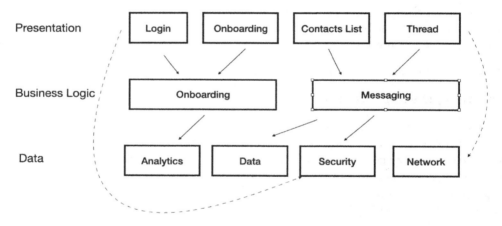

Figure 15.2: Selective closed layer

Figure 15.2 shows the same three layers we discussed earlier – presentation, business logic, and data. However, this time, we break the layer into different components. In addition, we display the various communication paths between the multiple components. For example, the login UI component communicates with the onboarding logic part of the app, and the messaging logic part communicates with the data and security components. Even though most communication goes through the business logic layer, we see some exceptions. For example, the following exceptions might apply:

- The Login UI component approaches the Security component directly, probably to understand the current authentication status
- The Thread UI component communicates with the Network components to present the network status in the UI

We can scratch our heads to find a way for these cases to go through the business logic layer; however, bypassing it and going directly to the data layer is perfectly acceptable in some cases.

The architecture we selected serves our project needs rather than vice versa. Yet we need to define a policy of bypassing the business logic layer since each exception, including those described in *Figure 15.2*, creates another coupling in our structure.

We discussed the three layers of architecture, but is it always three? Do we have more layers? Let's find out whether creating a more complex yet useful architecture is possible.

Adding more layers

Working with three layers is the sweet spot between simplicity and good separations. However, sometimes, the principle of separation of concerns still needs to be fulfilled in big projects. Even though it looks very straightforward to have one layer for the presentation and another one for the business logic, there are some dilemmas that need clarification.

Let's take, for example, two different components we may have in an iOS app – a user service and a payment service. Both are part of the app's business logic. When the user wants to make a purchase, we want to check their role using the user service and then go to the payment service to make the purchase. Right after the purchase, we want to navigate the user to a screen and show them that the payment was successful. So, we can see that we have a use case that involves incorporating different business logic services and coordinating different screens (*Figure 15.3*):

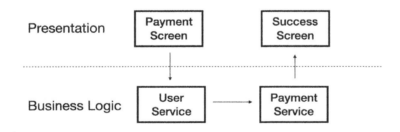

Figure 15.3: The payment use case, combining multiple components and layers

Do we need to manage that use case in the presentation logic, in the business logic, or half here and half there? Well, chances are that this logic is spread across components or centralized in one of the screen view models. Remember that a view model handles UI states rather than app logic in most cases.

The problem of handling use cases bundled with navigation is not new, and in apps that are more complex and require more flexibility, this needs to be taken into account when designing our app structure.

So, to separate our concerns, we can add another layer – the **Application** layer, which can handle a specific app use case (*Figure 15.4*):

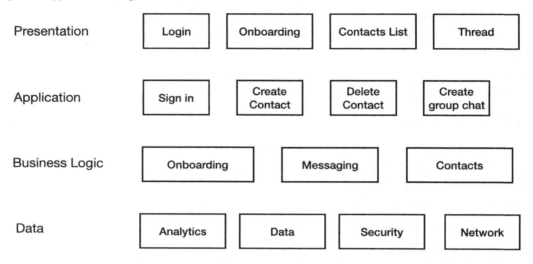

Figure 15.4: A four-layer architecture

Figure 15.4 shows an architecture design similar to *Figure 15.2*, this time with the Application layer. The Application layer has four use cases: Sign In, **Create Contact**, **Delete Contact**, and **Create group chat**. These use cases handle everything from calling functions in other components to navigation. The Application layer makes the business and presentation layers cleaner from specific app logic.

Do we have more layers we can add?

The Application layer coordinates multiple components to create an app-specific logic. We can implement the same concept on the bottom side of the architecture, between the business logic and the data layer.

For example, let's discuss a data sync process. Retrieving data from the network and storing it in the persistent store is a complex process that involves error handling and handling various edge cases. Is it part of the business logic or the data layer?

Data manipulation and **Create, Read, Update, and Delete** (**CRUD**) operations are also tasks that are unclear on which layer handles them.

So, to handle tasks that are not business logic but focus on accessing and managing data from various sources, we can add another layer called the **Data Access layer** (*Figure 15.5*):

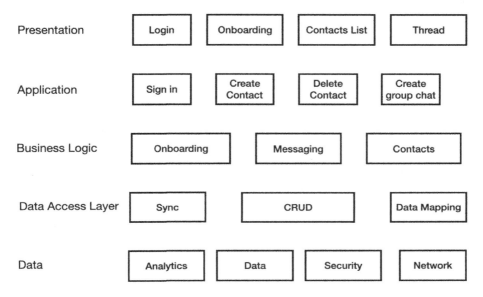

Figure 15.5: Data Access layer

Figure 15.5 shows our architecture from *Figure 15.4*, now with an additional layer – the Data Access layer, which handles sync service, CRUD operations, and data mapping, transforming data model objects into business logic entities.

Having more than three layers may sound too complex and imply over-engineering. However, this strategy ensures excellent separation of concerns between the different layers. The business logic doesn't involve data manipulation, and the presentation layer doesn't handle complex use cases. In medium and large apps, separating our project into four or five layers can pay off when our app gets bigger.

Separating into different layers is only one perspective we can consider in our project. I mentioned earlier that architectural patterns such as this act more like principles. The real power comes when we combine different patterns. Let's explore the modular architectural pattern.

Separating our project into modules

I mentioned that separating into layers is only one perspective we can look at in our project. However, what does it mean?

Let's take our messaging app as an example. The different layers represent different concerns: presentation, business logic, and data. Our app data flows through the layers from the UI to the data and back.

Another way to look at our app is through code units that encapsulate a set of functionalities or a specific business domain unit. We can call these code units **modules**.

Understanding the different considerations when working on modules

Separating our project into modules requires careful consideration, as this step is crucial for the app's structure over time. In a messaging app, for example, the modules can be a user authentication module, user profile module, contacts module, and messaging module. These modules reflect the app's different domains and the decision to divide our app is very flexible.

However, some key factors can help us decide:

- **Functionality and business domain**: We already mentioned this in the previous paragraph. Breaking down the app into core features can be an excellent start to understanding our project's different modules – logic, song player, reminders, onboarding, and more.
- **Reusability**: Grouping functionalities we use across different parts of the app is another way of understanding how to create a module. For example, if our app performs different HTTP requests, we may create a network module to manage all the API calls. Another example can be shared components – if we use the same button on different screens, that can be a sign that it should be part of a UI module that contains different reusable components.
- **Decoupling**: Our module should be decoupled from the other modules as much as possible. The level of interdependence the module has can define whether it was an excellent call to create it as a module. In addition, if it's possible to make a clear interface for the module, that's another indication it can be a good module.
- **Collaboration**: Imagine that several teams are working on our project. The fact that they can work without stepping on each other's toes signals good module separation. Note that the relevance of this rule remains the same, whether we are a team of one person or five teams with six developers each. The principle is what counts.

We must ask ourselves: can we create another app and use some of our modules in the new app, like using Lego bricks? Can we test each module separately? These questions give some sense of whether our modules are indeed independent or have tight coupling.

Organizing the code in our project

A few words about organizing our code into modules – modules are an abstract definition, as there's no official way of technically separating our code into modules. However, we can distinguish between two approaches – **physical** and **functional**:

- In the physical approach, we create our modules using a dedicated tool. CocoaPods, Swift Packages, and XCFrameworks, for example, provide a way of physically encapsulating our code into code units.
- In the functional approach, we do not use any specific tool but instead organize the code into folders. This simple approach is great for small projects or teams.

The primary consideration here is obvious: reusability and independence in the physical approach versus simplicity and flexibility in the functional approach. However, let's delve deeper and make this comparison more practical and relevant to our day-to-day work as iOS developers.

Creating a new project and understanding the different modules can be quite challenging. On the one hand, good planning is crucial for the success of our project development over time. On the other hand, it's impossible to predict how our project will evolve over the years. So, what we need at the beginning is flexibility. Therefore, starting with a functional approach, which involves creating modules by folders, might be the right approach for most projects.

As the project grows, the advantage of having flexibility in our modules can become its downfall. One of the great things about the physical approach is that we create clear boundaries between our modules by encapsulating our code into pods or packages. These boundaries prevent us from including external classes and types without handling the different dependencies correctly. They also force us to declare a clear interface for the module, as private and internal functions and classes are inaccessible from the outside. These restrictions are essential and valuable as the project evolves and the development team grows. The different pods or packages allow other teams to work on each module and build and test it separately. It even lets us share the same module between various projects.

So, now that we are convinced that modules are important, how does the idea of layers fit in? Do we have to choose between layers and modules? Or is it the same thing?

Let's try to put things in order.

Combining the multi-layer architecture with modules

We previously said that multi-layers and modules are more like architectural patterns or concepts. They are the guidelines for structuring our app and it is common practice to combine different concepts and patterns in our projects rather than stick with only one pattern.

Let's take the app onboarding module, for example (*Figure 15.6*):

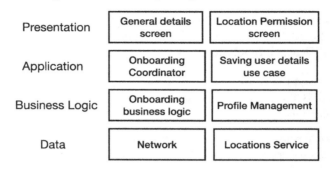

Figure 15.6: The onboarding module

Figure 15.6 shows the onboarding module structure divided into four different layers. One way to combine modular and multi-layer architectures is to create a matrix structure separating each module into different layers. In this case, the onboarding module has a presentation, coordinator, business logic, and data.

The other case involves layers built from several modules. In this case, each layer is a business unit built from several modules.

Let's compare both architectures:

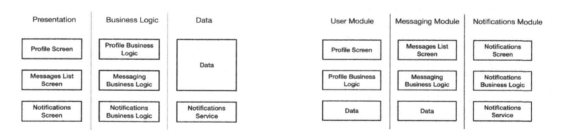

Figure 15.7: Comparing the two approaches of combining modules and layers

Figure 15.7 shows the two approaches we discussed side by side. At first glance, these two approaches look similar, just from different points of view. However, they represent two different project requirements and dramatically influence scalability, independence, and coupling.

Let's take, for example, the data modules. On the right side (**Modules comprising multiple layers**), we can see that each module has its data module. However, on the left side (**Layers built from multiple modules**), we have one data module that can serve various screens and business units. The same goes for more data layer modules such as analytics, network, and security.

When we consider it, for a module to be truly independent, it needs to contain all the layers and services. This also means that we'll have to duplicate some of the code in some cases.

As always, we have a tradeoff between encapsulation and independence versus centralized logic and consistency. Therefore, in practice, we must balance that, create a hybrid approach, and combine elements of both methods (*Figure 15.8*):

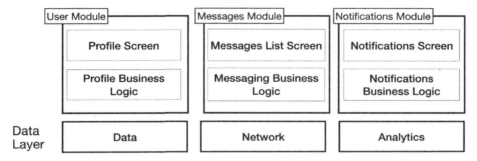

Figure 15.8: Combining layers built from modules and modules comprising multiple layers

Figure 15.8 presents the hybrid approach we discussed. Notice that we have several modules – **User**, **Messages**, and **Notifications** – each containing screens and business logic. However, the data, network, and analytics services are shared across the different modules.

The hybrid approach means that the different modules are only partially independent. On the other hand, it expresses a nice balance between reusability and encapsulation.

We can take it even further and share more logic, utilities, and UI components.

The multi-layer and module architectures are straightforward for most developers. They represent a logical way of examining apps and projects – either by levels of concerns, domains, or both.

Can we take a different approach to architecture? Let's try looking at our app differently – using **hexagonal architecture**.

Building hexagonal architecture

Multi-layer architecture describes an app as data flowing through different layers of concerns. The module architecture describes the app as different modules communicating with each other.

To go over an architecture with a different approach, we discuss the meaning of an app. What is an app? Is it the screens? Is it the logic?

In hexagonal architecture, we consider the business logic to be the heart and soul of the app. Let's take our messaging app example. The app's core is the messaging logic, the way we authenticate, and the different data models. We call that part of the business logic the **domain model**.

And what about the different screens, core data, and network layers? In the hexagonal architecture, these app parts are not at the core. The UI screens seem like the domain model's clients, and the network and core data parts provide services to the domain layer.

Look at *Figure 15.9*:

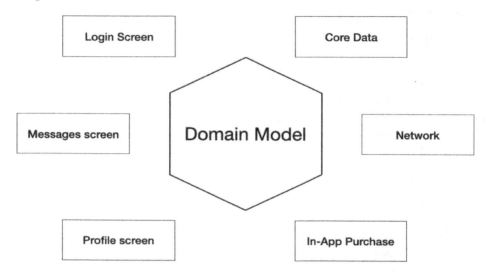

Figure 15.9: The domain model and its clients and services

Figure 15.9 shows the domain model at the center while its clients and services surround it.

Our next concept related to the hexagonal architecture is how the different actors connect to the domain model. These actors connect using ports and adapters.

Learning the concept of ports and adapters

Think of our app as a computer system. The computer has its motherboard, CPU, GPU, and memory. We can connect external input peripherals such as a keyboard, a mouse, or a trackpad. We can also connect output devices such as a display, a speaker, or a printer.

We know how computers are built – if we need to point out what is considered to be the heart of the computer, it wouldn't be the printer or the displayer, but rather its motherboard and CPU. However, can we connect any device we want to the computer?

To do that, we need two things:

- A port on the computer that allows us to connect devices; for example, UBC or HDMI
- A driver installed on the device that knows how to work with the port and the interface the computer requires

Each keyboard, printer, or mouse has a plug that fits the computer port and a driver that implements some protocol that allows this device to communicate with the computer. In general, we can connect any device we want as long as it conforms to the protocol the computer demands.

When we return to the hexagonal architecture, we can consider the domain model as the computer itself and the network or the UI as the printer and keyboard. In addition, we have two more terms – port and adapter:

- **Port**: This is an entry or exit point to/from the domain. In Swift, we use protocols to describe a port.
- **Adapter**: When a particular class wants to connect to a port, it needs to implement the port protocol.

Most iOS developers are familiar with the concept of a port and adapter. Eventually, this will be another way to decouple two elements using a protocol. However, in hexagonal architecture, all the elements that want to communicate with the domain model must use protocols.

There are two types of adapters – driving and driven. The distinction between them is fundamental to understanding the concept of hexagonal architecture.

Understanding the driving adapters

The driving adapters act as the entry point for the external world and are responsible for initiating any interaction with the domain model.

If we return to the computer example, adapters can be considered an external keyboard or mouse. We call them *driving* because they drive the app by invoking its use cases.

The most common example of driving adapters is the UI. A screen usually performs actions that drive our system to take meaningful action, such as logging in to the system, playing music, or fetching data from the network or the local persistent store.

However, driving adapters are not limited to UI screens. We can consider notification centers, app/scene delegates, location services, and universal links as driving adapters.

The driving adapter depends on the domain model and communicates with it only with protocols (ports).

Now, let's understand what the driven adapter is.

Understanding the driven adapter

The domain model uses the driven adapters to communicate with external systems or services, such as the network, persistent storage, or third-party services.

In the computer example, we can look at the driven adapter as an external display or a printer.

We can consider the whole hexagonal architecture as an I/O system – the driving adapters are the input devices, and the driven adapters are the output devices, performing updates to the local storage or executing API calls.

Let's look at our architecture now that we understand what ports, driving adapters, and driven adapters are (*Figure 15.10*):

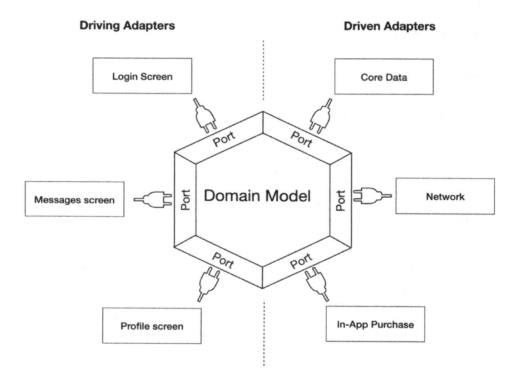

Figure 15.10: The complete hexagonal architecture

Figure 15.10 shows the different adapters, divided into driving and driven. It also shows that we need a port to access the domain model.

At this point, we have discussed the hexagonal architecture mostly in theory. Let's examine some examples of how to implement this concept in practice.

Implementing the hexagonal architecture in practice

Let's demonstrate the hexagonal architecture using a simple flow such as login.

Defining different ports

We start by defining the different ports. The first port is the login use case itself:

```
protocol LoginUseCaseProtocol {
    func login(username: String,
               password: String,
               completion: @escaping (Result<User, Error>)
                 -> Void)
}
```

The `LoginUseCaseProtocol` protocol defines how the driving adapter or app UI communicates with the app code, which is the domain model.

Our second port is one that we use to connect to a driven adapter, such as the network service:

```
enum NetworkRequestType{
    case login
}
protocol NetworkServiceProtocol {

    func performRequest(requestType: NetworkRequestType,
                        params: [String: Any],
                        completion: @escaping (Result<User,
                        Error>) -> Void )
}
```

The `NetworkServiceProtocol` protocol helps the domain model to communicate with external services such as network services.

Creating a login use case

Now that we have defined the different ports, we can create the login use case that sits at the heart of the domain model:

```
class LoginUseCase: LoginUseCaseProtocol {
    let authService: NetworkServiceProtocol

    init(authService: NetworkServiceProtocol) {
        self.authService = authService
    }

    func login(username: String, password: String,
      completion: @escaping (Result<User, any Error>) ->
      Void) {
        authService.performRequest(requestType: .login,
```

```
                              params: ["username" : username,
                                      "password" : password],
                              completion: completion)
    }
}
```

The `LoginUseCase` class implements the `LoginUseCaseProtocol` protocol, one of the ports we discussed earlier. It also uses the `NetworkServiceProtocol` protocol as a dependency. At this point, we have the login logic wrapped with a protocol and also communicate with the network service using a protocol. This means that the domain logic of our app is completely decoupled from the driving or driven adapter we may have, which is exactly what we wanted.

Creating a network service

Now, let's create a network service:

```
class NetworkService { }

extension NetworkService: NetworkServiceProtocol {
    func performRequest(requestType: NetworkRequestType,
                        params: [String : Any],
                        completion: @escaping (Result<User, any Error>) -> Void) {
        // implementation needed
    }
}
```

The `NetworkService` class implements the `NetworkServiceProtocol` protocol so that we can use it as a domain model dependency.

Creating a login screen

Now, let's turn to the driving adapter and create a login screen:

```
import SwiftUI

struct LoginView: View {
    
    @State var username: String = ""
    @State var password: String = ""
    
    let loginUseCase: LoginUseCaseProtocol
    
    var body: some View {
        VStack {
```

```
            TextField("Username", text: $username)
            SecureField("Password", text: $password)
            Button("Login") {
                loginUseCase.login(username: username,
                   password: password) { result in
                    // handle result
                }
            }
        }
        .padding()
```

In this example, we create a simple login screen (username and password) that uses its protocol to work with the login use case. If we need advanced state management, we can do that using a view model.

Connecting everything together

Now, all we need to do is to connect everything together:

```
@main
struct HexagonalAppApp: App {
    var body: some Scene {
        WindowGroup {
            let networkService = NetworkService()
            let loginUserCase =
               LoginUseCase(networkService: networkService)
            LoginView(loginUseCase: loginUserCase)
        }
    }
}
```

In the app initialization, we first create the driven adapters (the `NetworkService` class), inject them into the domain model (the login use case), and then inject the domain model into the driving adapter (the `LoginView` structure).

At first glance, it appeared we'd created too many protocols and used more dependency injection than usual. While it's true that this is the cost of using architecture such as Hexagonal, let's examine the benefits here:

- The different concerns are very clear. We understand exactly what the app's core logic is, what the external services are, and what the client of these modules is.
- Maintaining each adapter or core logic case is extremely easy since they are decoupled from each other and communicate only with a protocol. When we say maintain, we mean testing, refactoring, and bug fixes.

- Replacing parts in our app, such as services or use cases, becomes very easy. Let's try to remember apps or even systems that we have worked on. Imagine what it took to replace the network service, the persistent store, or even one screen.
- Adding more features and modules doesn't require significant changes to our project. Reusing existing code when adding new screens or use cases is easy.

Remember that, like the multi-layer and modular architectures, the hexagonal architecture provides a set of guidelines and principles for conducting a well-structured and maintainable project architecture.

So, how do these principles compare?

Comparing the different architectures

What is the best architecture we can use? Is there even right or wrong here? How do we digest all of that?

So, we saw how to combine modular and multi-layer architecture and emphasize each architecture's advantage. The same goes for the hexagonal architecture – let's pull out the different principles we've learned:

- Use protocols to decouple the communication with the external services
- Make the domain model the core of the app

These principles are relevant not only to hexagonal architecture but also to other architectures.

Let's try to compare the different architectures using several important metrics.

By separation of concerns

Separation of concerns is an important principle in project structuring, and all three architectures implement it well.

However, each separates the concerns in a slightly different way. For example, the *multi-layered architecture* separations are clear and straightforward, but they may lead to tight coupling if not carefully implemented.

On the other hand, in *modular architecture*, the separations are easy to maintain and scale because each module contains its own different layers and is self-contained. However, defining the distinct boundaries between the modules can be complex.

The *hexagonal architecture* focuses on separating the application core from external services. This approach is practical when adapting many external systems to the app. However, it requires a complex setup that can be overhead in small apps.

All these architectures have great separations of concerns because that's one of the most important principles in designing an architecture. However, each does that using a different approach, and the decision of the prominent architecture depends on the project requirements.

Let's see how to compare the different architectures in terms of testing.

By testing

Testing and separation of concerns principles are related to each other. The separation of concerns principle encourages the isolation of different classes and modules, which makes it simpler to write unit tests for a specific part of our app. Separating our apps based on concerns also makes it easier to manage dependencies, a critical factor in testing. However, since each architecture performs the separation differently, it also influences the testing.

For example, in a *multi-layer architecture*, it becomes easier to test each layer independently. We can perform core data or specific business logic testing in a simple way. However, if we want to write integration tests (tests that involve working with multiple components), the multi-layer architecture makes it much more complex because of the dependencies between the layers.

However, writing integration tests is one advantage of *modular architecture*, as the different interfaces within the module are well-defined. On the other hand, trying to write a unit test for a specific app layer can become much more complex now.

In the *hexagonal architecture*, we work with adapters and ports. This means loose coupling and many protocols with external services, which allows us to mock external services easily and easily test the application core.

To summarize, testing is a big topic in development, and each of the architectures supports it very well. To understand how the choice of architecture affects testing, we need to ask ourselves what the core unit we want to test is – is it a module, a layer, or the application core? Also, are integration tests important to us? Answering these questions can help us understand which architecture fits our project better.

What about maintenance and scalability? Let's see now.

By maintenance and scalability

Before we see how each architecture stands out regarding maintenance and scalability, let's understand exactly what it means. Maintenance is ongoing to keep our project aligned with the changing requirements. This involves fixing bugs, making new features and improvements, refactoring, and optimizing. Scalability describes our ability to increase the number of features without redesigning our project. In general, a well-maintained project is often considered to be scalable. However, like testing and separations, each architecture has a different approach.

The multi-layer architecture is great for medium-sized projects. Due to the tight coupling between the layers, maintaining a clear multi-layer architecture over time can be challenging in large projects. Modular architecture is considered to be highly scalable in big projects, as there are clear boundaries and independence between different business units. However, defining these units can be a challenge in the early stages. The hexagonal architecture is excellent for scaling – the clear domain separations help add more services to the project and test them over time. However, the maintenance can be overhead due to the many adapters we need to manage.

Each of the architectures is suitable for a different scale of project and requirements. Medium projects may work well with multi-layer architecture compared to modular, and hexagonal architecture can be great for large projects with one application core that can increase over the years.

Let's try to summarize by comparing the different architectures with different aspects:

Aspect	Multi-layer	Modular	Hexagonal
Separation of concerns	Clear, hierarchical layers (UI, logic, data); can become less flexible with dependencies	Independent, strong separation, flexible interfaces	Clear separation from external systems; core logic isolated via ports and adapters
Testing	Easy within the layers, complex between layers	Simple to test a single module, and the same goes for integration tests within the module	Core logic is very testable; easy to mock adapters
Maintenance	Can be challenging due to tight coupling	Easier due to modular; minimal impact across modules	Easy due to isolation from external changes
Scalability	Limited by layer interactions	Highly scalable as modules can scale independently	Scales by adding new adapters; core remains stable

This table can give us a sense of the different architectures' performance in different aspects. There are no scores here! We need to choose and mix the architectural concepts based on our needs.

Summary

Focusing on the correct architecture is a strategic decision that influences our project over time. If you feel confused about what fits your app, that's natural. Remember that the right thing to do is to look at the different architectures as different principles – we should combine the best from all worlds in a way that suits our project requirements.

In this chapter, we learned about the importance of architecture and what exactly it means. We also compared the different architectures – multi-layer, modular, and hexagonal. By now, you should be able to design your app's different components to help you scale, maintain, and test it over time.

It's not a coincidence that the book's last chapter discusses architecture. In a way, architecture ties together everything we've learned, providing a structure that allows all the elements to work together harmoniously. In addition, the app architecture is the infrastructure where we implement all the concepts we've learned throughout the book. Our journey ends; this is a good chance to start experiencing all the advanced iOS capabilities. Good luck!

Index

Symbols

1D bar marks
  adding 193-195
@Bindable property
  used, for binding objects 63-65
@Observable macro
  adding 52, 53
  computed variables, observing 56-58
  properties from observation, excluding with @ObservationIgnored 55, 56
  working 53, 54
@ObservationIgnored property
  used, for excluding properties from observation 55, 56
@Property attribute 313
@Query macro
  data, sorting 40, 41
  query, filtering 39, 40
  viewing, with data connection 39

A

Abstract Syntax Tree
  building 222, 223
  investigating 224-226
adapter 369
advanced animations
  keyframe animation, executing 129
  performing 125
  transitions, performing 125
advanced animations, transitions
  built-in transitions, implementing 125-127
  custom transition, creating 127-129
advanced operators
  exploring 262
AI
  basics 272
  versus machine learning 272
AI revolution 19
animations
  animation view modifier, using 121, 122
  need for 120
  performing 121
  withAnimation function, using 122-124
  with spring animations 124
animation view modifier
  using 121, 122
AnyPublisher
  working with 261, 262
app data flow
  controlling 359-361
app entities
  properties, selecting 317, 318
  used, for formalizing content 312

Index

AppEntity
   conforming to 312, 313
app flows
   network calls, integrating within 175
App Intents 301, 302
   chaining 315, 316
   confirmation and conditions, adding 310, 311
   creating 302, 303
   custom view, returning 307-309
   integrating, to other intents 316
   multiple result types 309, 310
   parameter, adding to 305-307
   reference link 322
   running, with Shortcuts app 303, 304
   used, for working with Apple Intelligence 320
Apple Intelligence 274
Apple platforms
   testing history 329, 330
Apple's TipKit framework 140, 141
   actions, adding 152-154
   popover tip, adding 146, 147
   tip ID, defining 148
   Tip model, defining 144, 145
   tips, adding 142, 143
   tips appearance 149
   tips, customizing 149
   tips, dismissing 147
Apple's TipKit framework, tips appearance
   tip properties, modifying 149, 150
   TipViewStyle, using 150, 151
app shortcut
   creating 304, 305
architecture 357
   importance 355

architectures comparison 374
   by maintenance and scalability 375, 376
   by separation of concerns 374
   by testing 375
AreaMark chart
   creating 202-204
arguments
   working with 338-340
AssistantEntity
   creating 323-325
Assistant Schema
   exploring 320-323
asymmetric transition 126

B

Bag of Words (BoW) 276
BarMark chart
   1D bar marks, adding 193-195
   creating 190
   interval bar charts, adding 195, 196
   Stacked Marks, adding 191
basic app architecture 165
   business logic 164
   data layer 164
   UI layer 164
basic HTTP request methods
   DELETE method 165
   GET method 165
   POST method 165
   PUT method 165
BERT algorithm 293
binding 63
built-in ML frameworks
   audio classification, with Sound Analysis framework 282-284
   exploring 274
   image analysis, with Vision framework 279

semantic search, performing with Core Spotlight 284
text interpretation, with NLP 275

C

Catmull-Rom splines 131
ChartProxy
  used, for allowing interaction 208
charts
  creating 190
  need for 188
  overlay, adding to 209, 210
  used, for visualizing functions 206-208
ChatGPT 272
closest data point
  finding, to user's touch 211, 212
CloudKit 35
C macros 218
Combine
  advantages 184
  exploring 182-185
  need for, using 246, 247
  operators, connecting 249, 250
  publisher, working with 247, 248
  subscriber, setting up 248, 249
  with examples 264
Combine components 250
combineLatest
  used, for combining multiple values 263, 264
Combine stream 249
computed variables
  observing 56-58
Conditional Random Field 293
content
  formalizing, with app entities 312
Context parameter 97

Continuous Integrations/Continuous Deployment (CI/CD) 349
control widget
  adding 114-117
Coordinator pattern
  calling, coordinator straight from view 81, 82
  CoordinatorView, adding 80, 81
  object, building 79, 80
  principles 77, 78
  working with 77
CoordinatorView
  adding 80, 81
Core Data 4, 22, 165
CoreML
  custom models, integrating with 288
CoreML framework 288
Core Spotlight framework 285
  indexing 286
  querying 287
  searchable items, creating 285, 286
  semantic search, performing with 284
Create ML 289, 291
  Spam Classifier model, building 291
CRUD operations 362
CubicKeyframe 131
custom data stores
  in Swift Data 6-9
custom models
  integrating with CoreML 288
custom operator
  creating 259-261
custom publisher
  connecting, with custom subscriber 258
  creating 251, 252
custom subscriber
  creating 255

receive(completion: Subscribers.
　　Completion<Never>) 257
receive(_ input: Int) -> Subscribers.
　　Demand 256, 257
receive(subscription: Subscription) 256
custom transition 127

D

Data Access layer 363
data connection
　to view @Query macro 39
data fetching
　with model context 36, 37
data flow 359
data manipulating
　with model context 36, 37
data migration 42
　migration plan, connecting to container 46
　migration stages and plan, creating 44-46
　process, learning 42, 43
　version schema, creating 43, 44
data models
　using, to trigger navigation 71, 73
data store
　building 173-175
decision tree algorithm 273
delta updates 180
demand-driven model 248
display frequency
　customizing 161
　max display count, setting 161
　setting 161
donations 159
driven adapter 369, 370
driving adapters 369

E

endpoints
　issues 179, 180
end-to-end testing 330
environment variables
　adding, by key 61-63
　adding, by type 59-61
　working with 59
error handling
　implementing 167-170
events rule 157-159

F

floating tab bar
　adding 11-13
forms
　validating 267-269
Foundation framework 166
four-layer architecture 362
full data sync with delta updates 180, 181
　consideration 181, 182
functions
　visualizing, with charts 206-208

G

Gemini 273
getTimeline() function 97
group container 144

H

hexagonal architecture
　adapters 368
　building 367
　domain model 367

driven adapters 369, 370
driving adapters 369
implementing 370
login screen, creating 372, 373
login use case, creating 371, 372
network service, creating 372
ports 368
ports, defining 371
History API 6
HTTP request
basic HTTP request methods 165
handling 165
response handling 167
URLSession class 166, 167

I

image analysis
Vision framework, using 279
working 279, 280
incremental loading technique 178
factors, considering 179
inline tip 142
Intermediate Representation Generation (IRGen) 220
interval bar charts
adding 195, 196
iOS development
structure 356
iOS SDK
closures 246
delegates 246
Key-Value Observing (KVO) 246
notifications 246

J

just-in-time fetching technique 175, 176

K

keyframe animation 129-132
components 129
duration, handling 132, 133
keyframe types
CubicKeyframe 131
LinearKeyframe 131
MoveKeyframe 131
SpringKeyframe 131
K-Nearest Neighbors (KNN) 272

L

Large Language Model (LLM) 273
Left-to-Right (LTR) layout view 137
lemmatization 285
LinearKeyframe 131
LineMark charts
creating 196-200
Lists 142
List type 208
long-term influence 357
loop unrolling 218
Low-Level Virtual Machine (LLVM) 220

M

machine learning (ML) 274
basics 272
versus AI 272
macro 155, 218
benefits 218
compiler plugin, adding 236, 237
declaring 230
expansion function 233-236
implementing 232
running, with client 237

tests, adding 240-242
usage example 218
macro roles 230
  attached 230, 231
  freestanding 230, 231
macros errors
  handling 238-240
Mark 190
ML model
  delving into 273
  training 273
mobile networking 164
ModelContainer
  connecting, with modelContainer modifier 34
  exploring 33
  ModelConfiguration, working with 35, 36
  setting up 33, 34
model context
  objects, fetching 38
  objects, saving 37, 38
  used, for data fetching 36, 37
  used, for data manipulating 36, 37
modules 364
  considerations 364
MoveKeyframe 131
multi-layer architecture 358
  combining, with modules 365-367
  layers, adding 361
multiple sources
  searches, performing from 265-267
multiple values
  combining, using combineLatest 263, 264

N

NaturalLanguage API
  text classification 276, 277
  using 276
  word tagging 277-279
Natural Language Processing (NLP)
  feature extraction 275
  modeling 276
  preprocessing 275
  used, for text interpretation 275
  working 275
navigationDestination view modifier
  used, for separating navigation destination 69, 70
NavigationPath
  used, for working with different types of data 75-77
NavigationSplitView
  creating 83-85
  moving, to three columns 85-87
  used, for navigating with columns 82
NavigationStack
  components 69
  exploring 68, 69
NavigationStack, components
  data models, using to trigger navigation 71, 73
  navigation destination, separating with navigationDestination view modifier 69, 70
  responding, to path variable 73, 74
network calls
  integrating, within app flows 175
networking
  exploring 182-185
network response
  deserializing 170-172

O

objects
  binding, with @Bindable property 63, 64
  data, connecting to view @Query macro 39
  fetching 38
  fetching, with FetchDescriptor 38
  #Index macro for performance, adding 41
Observable
  migrating to 65
ObservableObject protocol 50, 51
Observation framework 67
Open a task intent
  creating 314, 315
operators
  connecting 249, 250
  custom operator, creating 259-261
  working with 258, 259
  working, with AnyPublisher 261, 262
overfitting 292
overlay
  adding, to chart 209, 210

P

pagination loading 179
parametrized testing 340
path variable
  responding to 73, 74
pipeline 249
plot 206
Plottable protocol
  conforming to 213, 214
PointMark chart
  creating 204, 205
policy parameter 97
popover tip 142, 143
ports 369

project
  code, organizing 365
  separating, into layers 358
  separating, into modules 364
project architecture 356
protection level 288
protocols
  used, for performing mocking 352, 353
Provider function 97
publisher
  working with 247, 248
Publisher Operators
  reference link 258
pure functions 351

R

read-through cache technique 176, 177
response
  data store, building 173-175
  error handling, implementing 167-170
  handling 167
  network response, deserializing 170-172
Right-to-Left (RTL) layout view 137

S

scatterplot chart 204
Schemes 341
  setting up 349, 350
scroll views
  controlling 13
  handling, in SwiftUI 13
  item's visibility, observing 15
  position, observing 13, 14
searches
  performing, from multiple sources 265-267

SectorMark chart
  creating 200-202
semantic search 284, 285
  implementing 287, 288
  performing, with Core Spotlight 284
SenTestingKit 329
separation of concerns (SoC) 69, 374
SF Symbols 133
  animating 133-135
  apps, localizing 137, 138
  reference link 134
  symbol colors, modifying 135, 136
Shortcuts app
  used, for running App Intents 303, 304
singleton 166
smoke tests 337
Sound Analysis framework
  used, for classifying audio 282-284
Spam Classifier model
  building 291
  data, preparing 291-293
  deploying 298
  testing data 292
  training dataset 292
  training, preparing 293-297
  using, with Core ML 298, 299
  validation data 292
SpringKeyframe 131
stacked bar
  adding, to existing chart 191, 192
statement 224
state rule 154
  display rule, defining 156
  parameter, adding 155, 156
  parameter, connecting 157
StructInit macro
  defining 231, 232

StructInit struct
  declaring 232, 233
structs 332
Subjects
  PassthroughSubject 252, 253
  state, preserving with CurrentValueSubject Subject 253-255
  working with 252
subscriber
  setting up 248, 249
subtitles for the deaf or hard of hearing (SDH) 284
sub-views
  positioning, from another view 17-19
Swift Charts framework 187-189
Swift code
  generating 227
  parsing 220
SwiftData 21
  background 22
Swift Data improvements 5, 6
  custom data stores 6-9
  History API 6
  unique value 6
SwiftData model
  @Attribute macro, adding 30-32
  defining 23
  deletion rules 27, 28
  inverse relationship, defining 28-30
  @Model macro, expanding 23, 25
  relationship, adding 26, 27
  transient attribute 32, 33
Swift macros 217, 218
  creating 227-229
  package structure, examining 229
Swift Macros 322
Swift package 220

Swift Package Manager 221
  reference link 221
Swift Playgrounds 224
SwiftSyntax 219, 220
  Intermediate Representation
    Generation (IRGen) 220
  LLVM linking 220
  parse and abstract syntax tree (AST) 220
  semantic analysis (sema) 220
  setting up 220-222
  Swift Intermediate Language
    Generation (SILGen) 220
Swift Testing 5
  basic test, adding 331-333
  exploring 330, 331
  names, providing to test functions 334
  test functions, tagging 336, 337
  tests, enabling and disabling 334, 336
  working, with arguments 338-340
SwiftUI 4
  navigation, challenge 68
SwiftUI animations
  concept 120
SwiftUI observation system 50
  confirming, to ObservableObject
    protocol 50, 51
  problem, with current observation
    situation 51

T

tag 277, 336
testable code
  code based on concerns, separating 351, 352
  protocols, used for performing
    mocking 352, 353
  pure functions, writing 351
  writing, tips 350

test functions
  grouping, into test suites 342-344
  names, providing to 334
  tagging 336, 337
testing
  history, in Apple platforms 329
  significance 328
testing structure 341
testing suite 341
test plans 341
  adding 345, 346
  building 344
  configuring 347, 348
  customizing 346
  development processes 345
tests
  managing 340
text rendering behavior
  modifying 16, 17
three-layer architecture 359
  business logic 359
  closed layer 360
  data layer 359
  of messaging app 360
  open layer 359
  presentation 359
timeline 91
TimelineEntry structure
  building 99, 100
  components 99
Timeline Provider for Widgets 95, 97
  TimelineEntry structure, building 99, 100
  timeline, generating 97-99
Timeline reload policy 97
Tip protocol 144
tips rules
  adding 154
  events rule 157-159

state rule 154
TipGroup class, using 159, 160
tokens 226, 277
to-many relationships 27
to-one relationships 27
Transferable protocol
  using, to pass entire entity 318-320
transient attribute 32

U

UIKit-based view state
  managing, in view model 264, 265
UIKit/SwiftUI 188
UI testing
  performing 330
unique value 6
Universally Unique Identifier (UUID) 312
URL Loading System 166
URLSession
  initializing, ways 166
  working with 166, 167
User Interface (UI) 357
user's gesture
  responding to 211

V

view model
  UIKit-based view state, managing 264, 265
View Style 150
Vision framework
  barcodes, detecting 280, 281
  capabilities 280
  detection capabilities 281, 282
  faces, detecting 281
  using, for image analysis 279
VStacks 142

W

widget 90
  adding 91-93
  configuring 93, 94
  key roles 90
  network updates 110, 111
  reloading, with WidgetCenter 108-110
  static widget, configuring 95
  Timeline Provider for Widgets 95
  working 91, 108
widget customization 104
  AppEntity, adding 105
  EntityQuery, building 105-107
  intent, adding 104
widget interaction 111
  interactive capabilities, adding 112
  screen, opening with links 111, 112
widget interaction, interactive capabilities
  App Intents, using 113
  data change, performing 113, 114
  widget UI, updating 114
WidgetKit 4
widget UI
  animations, adding 102-104
  AppEntity, using 107
  building 100, 101
  timeline entries, working with 101
  widget customization 104
withAnimation function
  using 122-124
word embedding 279

X

XCTest 4, 329

Y

You Only Look Once (YOLO) 272

Z

zip operator
  using 262, 263
zoom transition 9-11

packtpub.com

Subscribe to our online digital library for full access to over 7,000 books and videos, as well as industry leading tools to help you plan your personal development and advance your career. For more information, please visit our website.

Why subscribe?

- Spend less time learning and more time coding with practical eBooks and Videos from over 4,000 industry professionals
- Improve your learning with Skill Plans built especially for you
- Get a free eBook or video every month
- Fully searchable for easy access to vital information
- Copy and paste, print, and bookmark content

Did you know that Packt offers eBook versions of every book published, with PDF and ePub files available? You can upgrade to the eBook version at packtpub.com and as a print book customer, you are entitled to a discount on the eBook copy. Get in touch with us at customercare@packtpub.com for more details.

At www.packtpub.com, you can also read a collection of free technical articles, sign up for a range of free newsletters, and receive exclusive discounts and offers on Packt books and eBooks.

Other Books You May Enjoy

If you enjoyed this book, you may be interested in these other books by Packt:

An iOS Developer's Guide to SwiftUI

Michele Fadda

ISBN: 978-1-80181-362-4

- Get to grips with UI coding across Apple platforms using SwiftUI
- Build modern apps, delving into complex architecture and asynchronous programming
- Explore animations, graphics, and user gestures to build responsive UIs
- Respond to asynchronous events and store and share data the modern way
- Add advanced features by integrating SwiftUI and UIKit to enhance your apps
- Gain proficiency in testing and debugging SwiftUI applications

Kotlin Design Patterns and Best Practices

Alexey Soshin

ISBN: 978-1-80512-776-5

- Utilize functional programming and coroutines with the Arrow framework
- Use classical design patterns in the Kotlin programming language
- Scale your applications with reactive and concurrent design patterns
- Discover best practices in Kotlin and explore its new features
- Apply the key principles of functional programming to Kotlin
- Find out how to write idiomatic Kotlin code and learn which patterns to avoid
- Harness the power of Kotlin to design concurrent and reliable systems with ease
- Create an effective microservice with Kotlin and the Ktor framework

Packt is searching for authors like you

If you're interested in becoming an author for Packt, please visit `authors.packtpub.com` and apply today. We have worked with thousands of developers and tech professionals, just like you, to help them share their insight with the global tech community. You can make a general application, apply for a specific hot topic that we are recruiting an author for, or submit your own idea.

Share Your Thoughts

Now you've finished *Mastering iOS 18 Development,* we'd love to hear your thoughts! Scan the QR code below to go straight to the Amazon review page for this book and share your feedback or leave a review on the site that you purchased it from.

`https://packt.link/r/1835468101`

Your review is important to us and the tech community and will help us make sure we're delivering excellent quality content.

Download a free PDF copy of this book

Thanks for purchasing this book!

Do you like to read on the go but are unable to carry your print books everywhere?

Is your eBook purchase not compatible with the device of your choice?

Don't worry, now with every Packt book you get a DRM-free PDF version of that book at no cost.

Read anywhere, any place, on any device. Search, copy, and paste code from your favorite technical books directly into your application.

The perks don't stop there, you can get exclusive access to discounts, newsletters, and great free content in your inbox daily

Follow these simple steps to get the benefits:

1. Scan the QR code or visit the link below

```
https://packt.link/free-ebook/9781835468104
```

2. Submit your proof of purchase
3. That's it! We'll send your free PDF and other benefits to your email directly

Made in United States
North Haven, CT
19 December 2024

63117244R00228